Emerging Trends in Mechatronics

Series Editor

Aydin Azizi, Oxford, UK

Md. Mizanur Rahman · Farhan Mahbub ·
Rumana Tasnim · Rezwan Us Saleheen
Editors

Mechatronics

Fundamentals and Applications

Springer

Editors
Md. Mizanur Rahman
Department of Mechatronics Engineering
World University of Bangladesh
Dhaka, Bangladesh

Farhan Mahbub
Department of Mechatronics Engineering
World University of Bangladesh
Dhaka, Bangladesh

Rumana Tasnim
Department of Mechatronics Engineering
World University of Bangladesh
Dhaka, Bangladesh

Rezwan Us Saleheen
Department of Mechatronics Engineering
World University of Bangladesh
Dhaka, Bangladesh

ISSN 2731-4855 ISSN 2731-4863 (electronic)
Emerging Trends in Mechatronics
ISBN 978-981-97-7116-5 ISBN 978-981-97-7117-2 (eBook)
https://doi.org/10.1007/978-981-97-7117-2

This Springer imprint is published by the registered company Springer Nature Singapore Pte Ltd.
The registered company address is: 152 Beach Road, #21-01/04 Gateway East, Singapore 189721, Singapore

Preface

The term 'mechatronics' refers to a pivotal interface between mechanics and electronics. Fundamentally, it is a synergistic approach to engineering that blurs the traditional differences among mechanical, electrical, and computer engineering, leading to innovative solutions that optimally leverage the strengths of each discipline. Mechatronics engineers build inventive solutions that integrate mechanical components with electronic and software systems to create and develop useful and efficient products. Mechatronics engineering highlights the inherent strengths of several disciplines, including electrical engineering, electronics engineering, mechanical engineering, and computer engineering, and intelligently consolidates these to design advanced electromechanical systems and processes. This diversified field of engineering aims to create efficient, reliable, and economical autonomous systems, bridging the gap across the subfields of this discipline. Mechatronics engineering contributes extensively to modern manufacturing, industrial automation, robotics, biomedical fields, automotive sectors, and so on. This book has ten chapters and emphasizes the relevant methodologies that encompass modeling, design approaches, and control of mechatronic systems. In the first chapter "Introduction to Mechatronics", the focus was on the historical evolution of mechatronics, demonstrating how mechatronics has become an essential element of modern engineering. It also has a focus on the foundational concepts of mechatronics, including analog and digital electronics, microprocessors, and microprocessor interfaces to electromechanical systems, as well as the interdisciplinary nature of the subject. The second chapter "Sensors and Actuators" illustrates various analog and digital sensors, such as optical encoders, Hall-effect sensors, potentiometers, gyroscopes, variable-inductance transducers, permanent magnet transducers, eddy-current transducers, variable-capacitance transducers, and piezoelectric transducers, among others. Undergraduate or graduate students in control engineering and mechatronics engineering programs, as well as researchers and practitioners seeking to deepen their quantitative understanding of physiological processes, are the target audience for the "Control System in Mechatronics Engineering" chapter. The fourth chapter "Robotics in Mechatronics Engineering" has provided a solid foundation for robotics within mechatronics and its diverse implications for various industries.

The fifth chapter "IOT Architecture and its Integration with Mechatronics" of this book discusses the foundation of IoT, focusing on the key concepts of designing IoT-based mechatronic solutions. It is also showcasing how it transforms today's rapidly changing industries by impeccably integrating smart devices and creating interconnected systems. The sixth chapter "Machine Learning and Artificial Intelligence" briefly discusses the challenges of implementing artificial intelligence (AI) and machine learning (ML) in mechatronics systems. The goal of this chapter is to develop an understanding of machine learning (ML) and AI techniques used in mechatronics engineering. The seventh chapter "Emerging Applications of Mechatronics" discussed the wide range of mechatronics-based technologies and all facets of intelligent applications of this technology. The eighth chapter "Ethical and Social Consideration in Mechatronics" provides insights into the vital interplay between technology and various sectors, highlighting the pivotal role of mechatronics engineering in shaping our rapidly evolving world. The ninth chapter "Case Studies: Application of Mechatronics Engineering" highlighted the significance of a professional robotics engineer's code of ethics, which acts as a roadmap for the ethical advancement of today's industry. It also emphasizes user diversity, including a range of viewpoints, requirements, and beliefs. This book explores the final chapter "Conclusion and Future Trends of Mechatronics", which focuses on state-of-the-art technologies like artificial intelligence, machine learning, and computational intelligence in mechatronics, illustrating various examples, recent advancements, and case studies from real-world implementations. Therefore, this book is going to serve as a reference tool for students, academics, practitioners, researchers, and industrial leaders in their respective fields.

Dhaka, Bangladesh

Md. Mizanur Rahman
Farhan Mahbub
Rumana Tasnim
Rezwan Us Saleheen

Contents

Introduction to Mechatronics

Ahmed Farhan, Protik Barua, Rezwan Us Saleheen, Rumana Tasnim, Md. Mizanur Rahman, and Mustafizur Rahman

Abstract The term 'mechatronics' refers to a pivotal interface between mechanics and electronics. It is fundamentally a synergistic approach to engineering, where the traditional differences among mechanical, electrical, and computer engineering are blurred, resulting in innovative solutions that best utilize the strengths of each discipline. Mechatronics engineers build inventive solutions that integrate mechanical components with electronic and software systems to create and develop useful and efficient products. This introductory chapter will illustrate how mechatronics acts as a multidisciplinary field that smoothly combines mechanics, electronics, and computation. This chapter will also highlight the historical evolution of mechatronics and demonstrate how mechatronics has become an essential element of modern engineering. This chapter will further focus on the foundational concepts of mechatronics, including analog and digital electronics, microprocessors, and microprocessor interfaces to electromechanical systems, as well as the interdisciplinary nature of the subject.

Keywords Mechatronics · Electromechanical · Artificial intelligence · Automation · Robotics

1 Introduction

Mechatronics is an integrative branch of engineering that combines physical components from electronics and mechanics with software engineering components. A mechatronic system's development calls for specialized multidisciplinary expertise from several engineering specialties. This is especially important when integrating a mechatronic system for full system testing, which combines all the parts and tests the use-case criteria supplied by the client [1]. Generally, mechatronics is a coherent step in developing modern engineering design. Computers, embedded computers,

A. Farhan (✉) · P. Barua · R. U. Saleheen · R. Tasnim · Md. M. Rahman · M. Rahman
Department of Mechatronics Engineering, World University of Bangladesh, Dhaka, Bangladesh
e-mail: farhan1@mte.wub.edu.bd

M. M. Rahman et al. (eds.), *Mechatronics*, Emerging Trends in Mechatronics,
https://doi.org/10.1007/978-981-97-7117-2_1

microcomputers, and other related software improvements and information technologies were introduced in the latter half of the twentieth century. These advanced technologies have made mechatronics indispensable. The field of mechatronics has a bright future ahead of it as the twenty-first century draws to a close because of expected advances in quantum computing, nano- and pico-systems, integrated bio-electro-mechanical systems, and other unexpected technologies [2].

Mechatronics has a number of definitions with respect to the views of authors. In 1969, a senior engineer at a Japanese company, Yaskawa, coined the term 'mechatronics' by combining the words 'mecha' (for mechanics) and 'tronics' (for electronics) [3–5]. The term was quickly adopted widely in the industry, and Yaskawa chose to give up his rights to it in 1982 to permit unrestricted use [6]. In other words, as goods and technologies progress, electronics will be incorporated into mechanisms in a more intimate and organic manner, making it more challenging to differentiate between the two [7]. Mechatronics, as defined by Fukada, Tomizuka, and Harashima, is the design and production of industrial products and processes that combine the synergistic elements of mechanical engineering, electronics, and intelligent computer control. This definition is frequently used and cited. The definition of the term 'mechatronics' has developed over the last thirty years. One way to characterize it would be as the integration of information technology (IT) and complex decision-making with physical systems in the creation, production, and management of industrial processes and products [8]. There are also some relative definitions of mechatronics, such as the following:

- Electronic control, mechanical systems, and precise engineering work harmoniously in mechatronics. It is the branch of research that exists at the intersection of informatics, automation, robotics, electronics, and mechanics [9].
- The diverse field of mechanics describes some skills essential in the modern and sophisticated automated manufacturing sector. Mechatronics experts can design more upfront, intelligent structures by combining the fields of electronics, mechanics, and computing. The anticipated intensification in manufacturing and automation requires mechatronics as the ultimate building block [10].

The knowledge and tools needed to generate controlled motions in a flexible manner are included in mechatronics [11]. The synergistic fusion of computer science, information technology, mechanical and electrical engineering, and control systems, along with numerical approaches for designing intelligent products, is known as mechatronics [12]. However, it is still not attainable or desirable to provide a particular definition of mechatronics because the area is still in its infancy and growing quickly [13]. Restricting and limiting the field would be exactly what is not desired right now.

2 Evolution of Mechatronics Engineering

In 1969, Tetsuro Mori introduced the term 'mechatronics' for the first time. He was a senior engineer at Yaskawa Electric Corporation in Japan. The phrase was coined by him to characterize the way industrial automation integrates mechanical and electrical engineering [14]. The development of mechatronics can be explained via some stages marked by the incorporation of different technologies into mechanical systems.

Prior to 1900, mechanical systems were purely based on mechanical principles. During that time, machinery and systems operated exclusively through mechanical means, including gears, levers, pulleys, and other mechanical components. During the Industrial Revolution, many systems were developed that incorporated steam engines, textile machinery, and early clocks. These systems made use of mechanics, hydraulics, and fundamental principles of thermodynamics. The 1920s saw the use of mechanical systems powered by electrical drives. During the 1920s, mechanical systems started integrating electrical drives, leading to a substantial enhancement in their efficiency and control. During this period, characterized by the growth of industry after World War I, factories were electrified, enabling the use of more efficient and adaptable technology. During this era, significant technological achievements were made, including the development of electric motors, generators, and fundamental electrical circuits. These breakthroughs paved the way for the creation of electrically powered conveyor belts, machine tools, and early electric locomotives. By approximately 1935, mechanical systems underwent advancements that incorporated automatic control mechanisms, enabling them to function with minimal human interaction. The pre-World War II era marked the introduction of feedback control systems, facilitated by advancements in control theory made by influential figures such as Norbert Wiener. Technological advancements such as feedback loops and early servomechanisms were widespread, leading to the development of automated weaving looms, early aircraft autopilots, and thermostats. By 1955, the combination of electronic control systems and sequential controllers allowed mechanical systems to attain enhanced precision and intricacy in their operations. The post-World War II era witnessed the initiation of the utilization of transistors and early computing in industrial applications. The main technological advancements were of transistors, early integrated circuits, and sequential logic controllers. These innovations ultimately resulted in the creation of numerical control (NC) machines, electronic relays, and automated assembly lines. The introduction of digital control systems in 1975 brought about a notable transition from analog to digital, enabling the continuous and advanced control of mechanical systems. The Digital Revolution, characterized by the prominent involvement of microprocessors and digital computers, occurred during this period. Technological advancements such as microprocessors, digital control systems, and programmable logic controllers (PLCs) have become crucial, leading to the development of inventions such as computer numerical control (CNC) machines, digital servos, and early robotic systems. In 1985, the area of mechatronics had developed, marked by the complete fusion of mechanical systems with

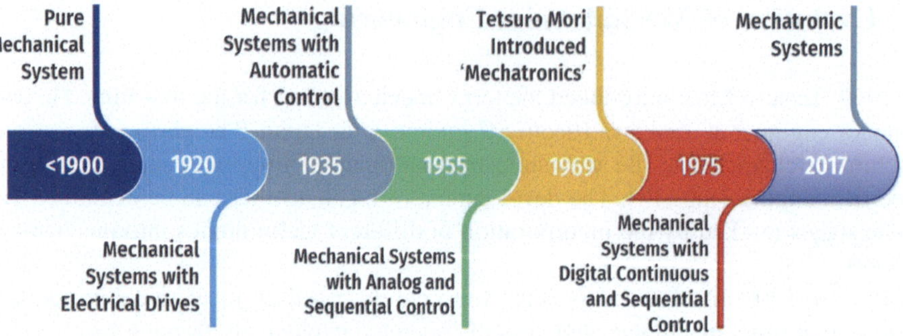

Fig. 1 Evolution of mechatronics system

electronics, sensors, and computer control. This period, which is a part of the Information Age, signified the official founding of mechatronics as a separate and identifiable discipline. Figure 1 illustrates the evolution of the mechatronics system. This figure highlights the progress of the mechatronics system from a pure mechanical system.

Advanced technology such as microcontrollers, sensors, actuators, and software engineering played a crucial role in the development of industrial robots, automated guided vehicles (AGVs), and advanced automobile systems such as anti-lock brake systems (ABS). Automated systems combining electrical control systems and mechanical components started to be implemented by factories. The trend toward machines that are more intelligent and adaptable began during this age. Across the world, universities and other organizations have launched robotics education and research programs. Mechatronics became a separate topic of study as the academic community realized that engineering needed to be approached from a multidisciplinary perspective [15–17].

The progression of mechatronics from purely mechanical systems to fully integrated mechatronic systems is a result of a sequence of technological developments. Each successive phase, starting from the implementation of electrical drives and automatic control and progressing to the incorporation of digital continuous control and advanced electronics, has relied on the progress made in the previous period, leading to the development of more intricate and proficient systems.

3 Mechatronics—A Combination of Mechanics, Electronics, Computation, and Control System

Mechatronics engineering is an interdisciplinary field that integrates components from several branches of engineering to develop unified and intelligent systems. Mechatronics engineering is the amalgamation of mechanics, electrical, control systems, and computer engineering. It creates novel solutions for a variety of applications by fusing the computational components of software, the electrical aspects of

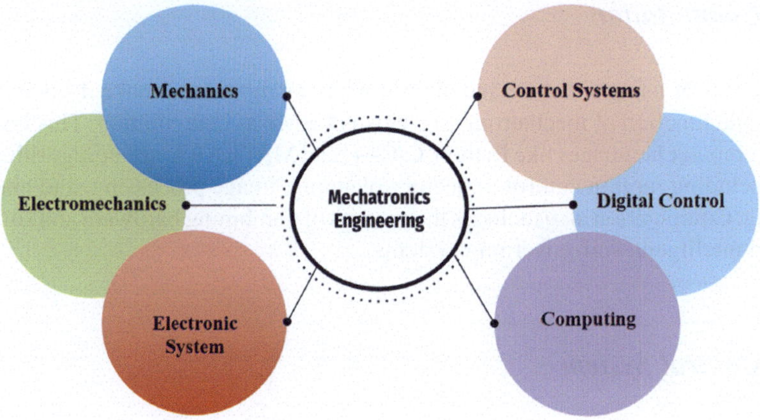

Fig. 2 Fundamental subfields of mechatronics engineering

electronics, and the physical aspects of mechanics. Figure 2 shows the fundamental subfields of mechatronics engineering.

3.1 Mechanics

The study of mechanical systems, including the creation, evaluation, and enhancement of mechanisms, structures, and materials, is the focus of mechatronics. To build the mechanical parts that make up the mechatronic systems' physical structure, one must comprehend ideas like kinematics, dynamics, and material strength.

3.2 Electronics

By giving mechatronic systems the ability to sense, process, and control physical quantities, electronics play a critical role. This covers the development and application of actuators to provide desired mechanical output as well as sensors to monitor variables like position, velocity, force, and temperature. The use of circuits, microcontrollers, and embedded systems for data acquisition, control algorithms, and signal processing is also included in the field of electronics [18].

3.3 Computation

The process of using software and algorithms to analyze data, make decisions, and control the function of mechatronic systems is known as computation. This involves using computer languages like Python, C/C++, and MATLAB to create algorithms for tasks including motion control, trajectory planning, image processing, and machine learning. Computation also includes the process of combining hardware and software to build intelligent, self-governing systems.

3.4 Control System

Control systems are essential in mechatronics engineering as they facilitate the integration of mechanical, electrical, and computer engineering to develop advanced and efficient systems. The primary purpose of the control system is to manage, instruct, or control the actions of other devices or systems through the utilization of control loops.

Mechatronics is the field that combines computation, electronics, mechanics, and control to create sophisticated systems and gadgets that are more flexible, functional, and efficient. Robotics, automation, automotive systems, medical devices, aerospace systems, and consumer electronics are a few examples of mechatronic applications. Overall, mechatronics acts as a link between conventional engineering specialties, enabling professionals to tackle difficult problems and develop creative solutions that draw on the advantages of several domains.

The design, analysis, and optimization of mechanical systems are the main focuses of mechatronics engineering. Numerous industries, including robotics, automation, automotive systems, manufacturing, aerospace, and biomedical engineering use mechatronics. The creation of more energy-efficient systems, the incorporation of artificial intelligence, component shrinking, and developments in collaborative robots are among the challenges and trends of the future [19].

4 Design Elements of a Mechatronics System

The Mechatronics system has several basic elements. Figure 3 demonstrates the design elements of a mechatronics system. This engineering field involves the design and analysis of sensors and actuators, signal processing and system modeling, data acquisition, computation, and logic circuit.

As these are interconnected, mechatronic systems are designed, analyzed, and optimized for numerous applications, such as robotics, automation, automotive systems, consumer electronics, healthcare, and various others.

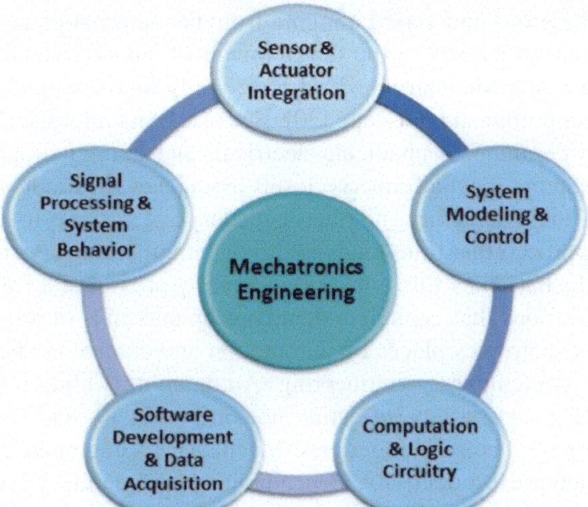

Fig. 3 Design elements of mechatronics systems

5 Mechatronics—An Essential Element of Modern Engineering

Within the broad field of contemporary engineering, mechatronics stands out as a crucial field that creatively combines mechanical, electrical, and computational components to design and create state-of-the-art products and systems. Mechatronics, which pushes innovation in many different industries and sectors, has become a multi-dimensional field. Fundamentally, mechatronics is the combination of the most recent developments in computing and electronics with the foundational ideas of mechanical engineering. This combination makes it possible to design intricate systems with previously unheard-of levels of automation, performance, and efficiency due to the cooperative efforts of mechanical parts, sensors, actuators, and computing devices. Mechatronics' multidisciplinary nature, which cuts across the borders of distinct engineering areas, is one of its distinguishing features. Mechatronics engineers are able to take on complex problems and come up with comprehensive solutions that meet the demands of contemporary society by utilizing knowledge and techniques from electrical engineering, mechanical engineering, control engineering, and computer science.

Mechatronics is a key player in determining the direction of technology in a number of industries, including consumer electronics, biomedical engineering, robotics, aerospace, and automobile manufacturing. Mechatronics is at the forefront of innovation, whether it is in the creation of sophisticated robotic systems that can operate autonomously in challenging conditions or in the planning of clever production techniques that maximize output and efficiency. Furthermore, beyond engineering, mechatronics influences a broad spectrum of sectors, including transportation, healthcare, agriculture, and entertainment. The uses of mechatronics are numerous and varied, ranging from the automation of medical diagnostic and treatment processes to the development of smart agricultural equipment for precision farming. Mechatronics' relevance only increases as we traverse the ever-changing technological landscape [20]. Because of its interdisciplinary character and capacity to combine mechanical, electrical, and computational components into cohesive systems, mechatronics is closely associated with contemporary engineering. Because of the complexity and interdependence of systems in modern engineering, a holistic approach that transcends traditional engineering specialties is frequently necessary. Mechatronics fills this requirement by providing a framework for inventing unique solutions that capitalize on advancements in a variety of industries. The focus that mechatronics places on automation and control is one of the main ways that it ties to contemporary engineering. Automation is ubiquitous in today's technologically advanced society, affecting not only industry and transportation, but also healthcare and consumer gadgets. Mechatronics engineers create automated systems that increase productivity, dependability, and safety by applying their knowledge of mechanical design, electronics, and programming. Additionally, mechatronics is essential to the creation of intelligent devices and systems. The need for smart goods and services that can adjust to shifting conditions and user requirements is rising because of the development of machine learning and artificial intelligence (AI). With the purpose of building intelligent systems with perception, decision-making, and action capabilities, mechatronics combines sensors, actuators, and computational algorithms. Furthermore, mechatronics makes advances in autonomous systems and robotics possible. The area of robotics is rapidly developing. It includes everything from autonomous cars and drones to industrial robots used in production. To ensure that these robotic systems can securely interact with things and people, navigate complicated surroundings, and carry out tasks with accuracy, mechatronics competence is a prerequisite for both developing and managing them. Mechatronics also has connections to other radical technologies like the Internet of Things (IoT), augmented reality (AR), and virtual reality (VR). Mechatronics makes it possible to create IoT-enabled goods and smart infrastructure that can gather, develop, and act upon data in real time by integrating sensors and connectivity into physical equipment. Through the incorporation of interactive interfaces and haptic feedback systems, mechatronics advances the creation of immersive VR and AR experiences. Because it offers the instruments, strategies, and approaches needed to tackle the intricate problems of the current technological environment, mechatronics is closely entwined with contemporary engineering. Mechatronics, with its multidisciplinary methodology, stimulates creativity, propels advancement, and molds the course of engineering and technology [6, 21].

6 Foundational Concepts of Mechatronics

Mechatronics' foundational notions cover a broad range of ideas and theories that serve as the cornerstone of this multidisciplinary science. The fundamental ideas of mechatronics involve an interdisciplinary approach that combines concepts from computer science, electrical engineering, mechanical engineering, and control systems. The ability of sensors and actuators to work together to sense and react to physical stimuli is the fundamental component of mechatronics. Control theory is essential for managing the behavior of electromechanical systems via feedback loops, and embedded systems are the foundation for managing and arranging the many parts of the system. Human factors and user-interface-design considerations guarantee intuitive human-mechatronic system interaction, while modeling and simulation techniques support system analysis and design optimization. For the creation of reliable and effective mechatronic systems with applications in robotics, automation, automotive systems, and other fields, the smooth integration of hardware and software components is essential [22].

6.1 Analog and Digital Electronics

The two main subfields of electronics that deal with the conversion, modification, and transmission of electrical signals are analog and digital electronics. Continuous signals, like voltage or current that fluctuate smoothly over time, are the main emphasis of analog electronics. It uses methods like amplification, filtering, and modulation together with parts including operational amplifiers, resistors, and capacitors. Applications like audio amplification, sensor interfacing, and communication systems demand accurate and real-time signal processing, which calls for analog electronics. Conversely, discrete signals in digital electronics are limited to two states: either high (logic 1) or low (logic 0). It makes use of parts like flip-flops, logic gates, and microcontrollers, as well as methods like sequential circuits, logic design, and binary arithmetic. Because digital electronics allow for precise and dependable manipulation of binary data, it is essential for jobs involving computing, data processing, and control. Digital computers, embedded systems, communication networks, and consumer devices all make substantial use of it. Digital electronics are robust, scalable, and easy to process and store signals. On the other hand, analog electronics are defined by their continuous nature and sensitivity to noise. These subfields are essential to contemporary electronics engineering and frequently work in tandem to design intricate systems.

6.1.1 Analog Electronics

- It deals with continuous smooth signals, such as current or voltage.

- Components include resistors, capacitors, and operational amplifiers.
- Techniques involve amplification, filtering, and modulation.
- Applications include audio amplification, sensor interfacing, and communication systems.

6.1.2 Digital Electronics

- It focuses on discrete signals with two states: low (logic 0) or high (logic 1).
- Components include logic gates, flip-flops, and microcontrollers.
- Techniques include binary arithmetic, logic design, and sequential circuits.
- Applications encompass digital computers, embedded systems, communication networks, and consumer electronics.

6.1.3 Microprocessors

Microprocessors investigate deeply into the basic ideas, construction, and uses of microprocessors in contemporary computing systems. From the earliest days of single-core processors to the multi-core and parallel processing architectures that are common in modern computer devices, it examines the development of microprocessor technology. The microprocessor architecture has important parts such as the control unit, registers, cache memory, instruction set architecture (ISA), and arithmetic logic unit (ALU). It addresses how to maximize the effectiveness and performance of contemporary microprocessors by utilizing pipelining, superscalar processing, and out-of-order execution. There are also some other microprocessor architectures, such as embedded processors, general-purpose CPUs, and application-specific integrated circuits (ASICs). It looks at how microprocessors function in a variety of computing systems, including embedded systems, Internet of Things (IoT) gadgets, high-performance computing clusters, and personal PCs and smartphones. To sum up, 'Microprocessors' is a priceless tool for anyone looking for a thorough grasp of microprocessor technology and how it has shaped the current computing environment. By combining theoretical ideas, real-world applications, and emerging trends, this chapter gives readers the information and abilities they need to successfully traverse the quickly changing realm of microprocessor technology.

6.1.4 Artificial Intelligence (AI) and Robotics

Artificial intelligence (AI) develops machines that typically require human intelligence and are capable of doing diverse tasks. These include the ability to solve issues, think through issues, recognize patterns, understand natural language, and learn through experience. Artificial intelligence is applied in robotics and mechatronics in a variety of ways, including artificial neural networks (ANN), fuzzy logic, genetic algorithms, machine learning, and evolutionary computing. Neural networks, deep learning, ensemble learning, reinforcement learning, and classical learning—both supervised and unsupervised—are some of the techniques used in machine learning. Intelligent control algorithms employ a range of statistical, probabilistic, and optimization techniques to assess massive data sets and identify meaningful patterns from actions taken [23].

In addition, fuzzy logic is an approach that is analogous to human cognitive processes and has many values. True or false are handled along with the absolute values of 0 and 1 in conventional Boolean and Aristotelian reasoning. Nevertheless, fuzzy logic operates differently. It deals with phenomena that lack absolute value and do not necessarily follow a linear function [24].

The processing and classification of data sets against neural networks are the main uses of classical learning and clusters, as well as reinforcement learning that may be used in the fields of autonomous control, logistics, and planning [25–27]. AI aims to mimic cognitive processes including language comprehension, learning, reasoning, and problem-solving, usually involving intricate algorithms and computer models. In order to create intelligent systems, this multidisciplinary area pulls from numerous subfields, including computer vision, robotics, NLP, machine learning, and deep learning. Each of these subfields offers a distinct viewpoint and methodology. Since its infancy in the middle of the twentieth century, AI has revolutionized numerous sectors and disciplines and is now widely used in contemporary society. AI has powered advances in healthcare, banking, education, entertainment, and more. AI models are computer frameworks that mimic the thought processes and behavior of humans, allowing machines to carry out tasks on their own and make data-driven judgments. These models cover a broad spectrum of methods and strategies designed for particular uses, ranging from straightforward rule-based systems to complex deep learning structures. Algorithms that process incoming data, identify relevant patterns or features, and produce predictions or actions as an output are the foundation of AI models. A subset of AI models called machine learning algorithms use past data or experience to get knowledge from past performance and gradually develop over time without the need for explicit programming. Artificial neural networks that mimic the structure and functionality of the human brain are used in deep learning, a type of machine learning, to solve complex issues like image recognition and NLP [28, 29].

6.1.5 Machine Learning

Machine learning (ML) is a branch of artificial intelligence that focuses on developing statistical models and algorithms that enable computers to learn from data and make decisions or predictions without the need for explicit programming. As ML algorithms are exposed to additional data over time, they iteratively improve their performance by learning from past examples or experiences [30]. The process typically involves the following:

- Acquiring pertinent information from a range of sources, including databases, sensors, and the Internet.

- Cleaning, transforming, and preparing the data for analysis, including tasks like scaling features, handling missing values, and encoding categorical variables.
- Choosing an algorithm or an applicable ML model based on the problem's nature, the category of data, and the anticipated outcome. Decision trees, linear regression, support vector machines (SVM), and NNs are the components of common ML techniques.
- Training the selected model with either unlabeled (for unsupervised learning) or labeled (for supervised learning) data. The model modifies its parameters during training in order to reduce the discrepancy between the values it predicts and the actual values.
- Depending on the particular job and goals, evaluate the trained model's performance using measures like precision, accuracy, recall, or F1 score.
- Integrating the trained model into an operational environment, where it can predict or make decisions in real time based on new data.

Machine learning has numerous applications. Some of them are as follows:

- Natural Language Processing: Text analysis, sentiment analysis, linguistic translation, and chatbots.
- Computer Vision: Facial recognition, object detection, image recognition, and medical imaging. Shaikat et al., (2023) proposed an attendance system based on real-time face recognition process. This system validates the real-time monitoring process [31].
- Predictive Analytics: Forecasting sales, predicting customer churn, risk assessment, and recommendation systems.
- Autonomous Vehicles: Self-driving cars, drones, and robotic systems.
- Healthcare: Disease identification, modified medicine, and drug innovation.
- Finance: Algorithmic trading, fraud detection, portfolio management, and credit scoring.

As data availability, computational power, and algorithmic advancements continue to evolve, machine learning plays an increasingly vital role in driving innovation and solving complex problems in various fields [32].

6.1.6 Neural Networks

Neural networks (NNs) are made up of layer-by-layer networked nodes. These nodes are known as neurons. Neurons generate output signals after processing input signals with an activation function. They are transmitted to other neurons within the network. The perceptron, a streamlined representation of a biological neuron, is the fundamental building block of a NN. Layers comprise an input layer, one or more than one hidden layer, and an output layer in the structure of a perceptron. Multiple hidden layers enable deep neural networks, also known as deep learning models, to recognize intricate patterns and representations from unprocessed data. They are particularly good at tasks like reinforcement learning, natural language processing

(NLP), and image and audio recognition. When a NN is trained, labeled data is fed into it (in the case of supervised learning), and the weights are iteratively adjusted using an optimization approach like gradient descent until the network generates accurate predictions. In order to train a neural network, two methods are utilized: back propagation, in which errors are computed and used to update the weights in the gradient's opposite direction, and forward propagation, in which input data is transmitted through the network to produce predictions. Neural networks have transformed numerous industries, such as computer vision, speech recognition, and autonomous driving. They have made advancements possible in fields including objective identification, language translation, game play, and image categorization. Neural networks are pushing the envelope of artificial intelligence and machine learning with each new development in technology, algorithms, and data availability.

6.1.7 Natural Language Processing (NLP)

NLP is a subfield of artificial intelligence (AI) that is concerned with how computers interact with human languages. It includes a range of strategies and tactics for deciphering, producing, and assessing natural language speech or text. This process is fundamentally the study and progress of models and procedures that allow computers to comprehend and deduce human language in a meaningful and practical way. This includes tasks such as the following:

6.1.8 Text Classification

To classify text documents into predefined groups or labels, such as spam recognition, sentiment investigation, or topic classification.

6.1.9 Named Entity Recognition

To identify and extract entitled objects from text, for example, names of people, organizations, locations, dates, and other entities.

6.1.10 Part-of-Speech Tagging

To assign grammatical tags to words in a sentence, for instance, nouns, adjectives, verbs, and adverbs.

6.1.11 Syntax and Parsing

To analyze the structure of sentences to understand their grammatical components and relationships between words.

6.1.12 Semantic Analysis

To extract the connotation of text and understand how words and concepts are related to each other. This analysis includes semantic similarity, word embedding, and semantic role labeling.

6.1.13 Machine Translation

To decode text using statistical models or neural networks.

6.1.14 Question Answering

To build systems that can understand and respond to natural language questions, such as virtual assistants or chatbots.

6.1.15 Text Generation

To generate coherent and contextually relevant text, such as through language modeling, text summarization, and dialogue generation [33].

6.2 Computer Vision

In contemporary surgical practice, a variety of surgical cameras are employed, offering a rich visual signal that helps surgeons make clinical judgments by evaluating the clinical site. Additionally, this visual signal can be utilized by AI techniques to assist in real time during actions and postoperatively for analytics. In order to extract important data and understandings from visual data, it examines the mathematical models, algorithms, and computational techniques used in the analysis and interpretation of digital photos and videos. In addition, computer vision has advanced subjects and new developments such as transfer learning, convolutional neural networks (CNNs), generative adversarial networks (GANs), recurrent neural networks (RNNs), and deep learning [34].

It looks at how these state-of-the-art methods have transformed a number of computer vision-related fields, leading to innovations in semantic segmentation, picture synthesis, scene understanding, and image identification.

7 Prospective of Mechatronics

Recently, the convolution of systems and expedients has increased significantly. It necessitates a system-level approach to mechatronics design. This method facilitates the easy execution of tests, the blending of mechanical and control design, and the reuse of algorithms inside the embedded delivery framework. By incorporating industry best practices with readily accessible technologies to expedite design, prototyping, and execution, this method significantly enhances the design process. It is possible for the engineers to create a more efficient procedure by dividing the design process into simultaneous threads. Previously, developing a control algorithm for a mechanical device required a team to wait for a workable prototype. To get started more quickly, a computer-generated prototype built from concept models and simulation data can be used by the engineers [35]. Engineers can reduce production time during deployment in a number of ways. Its goal is to reduce costs by creating a customized control system at the component level that can be installed on integrated COTS (commercial off-the-shelf systems) [36]. Customizable embedded systems, which offer the speedy deployment of COTS hardware and the versatility of a bespoke configuration, have become more popular in recent years. National Instruments launched its reconfigurable I/O platform approximately seven years ago. Users could add software algorithms to hardware using the graphical programming environment NI LabVIEW because of a programmable plug-in board for PCs that had an FPGA and integrated analog and digital I/O. From then on, this technology has developed and is now applied to industrial-scale system-level goods and durable embedded hardware [37].

In the future, software will be essential to mechatronics and embedded systems. Software tools give designers a system-wide perspective on various design elements and abstract the complexities of the latest technologies, allowing them to build high-performance systems capable of meeting future problems. Kuru et al. [38] proposed that the automation of everything symbolizes a shift in sophisticated mechatronic systems during the Industrial Revolution. The future of fear, the hidden life of data, and the ghost of computing are the three main developments in mechatronics [35].

- The ghost of computing is the result of electronics becoming smaller and less capable. As computer technology becomes more powerful and smaller, the focus shifts to the applications it can serve. Future mechatronic applications could benefit from a number of development projects that could expand the reach of mechatronic technologies to deeper and wider locations. The developing topics in mechatronics under this category include opto-mechatronics [39], biomedical engineering [35], nanoelectromechanical systems (NEMS) [40, 41], nano-mechanical systems (NMS) [35, 42], and nanotechnology [43, 44].

- 'Hidden Life of Data' refers to the use of big data and computer tools to understand human mental processes. This knowledge makes the data easier to utilize and gives it a sense of autonomy. Data science is a useful and essential skill [45–47]. Engineers, on a daily basis, are making amazing discoveries. The fields of robotics and data science are crucial ones to watch. However, they still have a long way to go before becoming self-sufficient; robots are becoming increasingly sophisticated and intelligent. For example, users cannot program a robot to grab everything that they want in order to operate it within a home [48]. Because of the advancements in artificial intelligence (AI), robots may soon be able to perform this type of simple activity on their own [35]. As technology advances, it is transforming how we live and conduct business. Researchers have created an AI-powered robot that can solve a task requiring children to place shaped pegs in the corresponding holes. Through a process known as machine learning [49], which is essentially automation through trial and error, the robot will be able to solve the beloved children's puzzle. To put the square peg into the square hole, researchers would guide the robot to figure out how to perform. More than fifty-three percent of global businesses use big data technologies [50]. Compared to only 17% in 2015, this is a significant increase. Organizational structure development and process optimization are being accomplished by an increasing number of businesses through the use of data science technology [51]. Data scientists now have a plethora of fascinating job options. This will work in tandem with other emerging technologies like blockchain, the Internet, artificial intelligence, automation, machine learning, and 5G [52–59] to create an incredible and highly advanced future.
- The concept behind 'the future of fear' is more theory-based than new technology. The goal of The Future of Fear is to silence people's worries about what technology will bring about in the future. It is human nature to provoke worries unrelated to the IoT or machines that relate to the future of technology. Technology is, after all, and always has been, a tool that we use to improve the quality of human life. Nowadays, advancements in electric technology are at the center of our entire world. As a result of innovative concepts we are able to implement, rather than the sheer size of technology we are creating, technological advancements continue to grow and change. The field of mechatronics offers opportunities for novel, ground-breaking concepts that have the potential to transform daily existence. Though technological advances and future developments have been anticipated in a variety of ways, technology companies and futurists are given resources to forecast how the world will change in ten or fifteen years, and further technological improvements are surely on the way [35].

8 Conclusion

Mechatronics is an interdisciplinary area that is expanding quickly and combines computer science, electrical engineering, and mechanical engineering to create intelligent systems that can communicate with their surroundings on their own. It is predicted to play a major role in many industries as the field develops, enhancing reliability, efficiency, and safety. Industrial companies are looking for candidates with multidisciplinary, comprehensive training who can handle challenges holistically and who have mastered mechanical, computer, automation, or electronic engineering to make creative improvements in products, systems, or processes that combine these disciplines. Because of this, mechatronics specialists are becoming more and more crucial. An overview of mechatronics, including its history, concepts, applications, difficulties, and design process, is highlighted in this chapter. Additionally, several potential developments and trends in the field of mechatronics have been covered in this chapter.

References

1. Axelsson A, Ossiansson V (2023) Viability of scrum elements in mechatronic system testing: an exploratory case study at Husqvarna AB
2. Ali S, Sabir S (Nov 2018) Developments in mechatronics engineering education and its current status in Pakistan. In: 2018 IEEE 10th international conference on engineering education (ICEED). IEEE, pp 69–74
3. Poduraev U (1999) Fundamentals of mechatronics. STANKIN
4. Kyura N, Oho H (1996) Mechatronics-an industrial perspective. IEEE/ASME Trans Mechatron 1(1):10–15
5. Mori T (1969) Yasakawa internal trademark application memo. Technical report 21131.01, Yasakawa Electric Corporation
6. Popovchenko M (2006) Introduction to mechatronics and mechatronics in real life. JASS 2006
7. Bishop RH (2017) Mechatronics: an introduction. CRC Press
8. Tomizuka M (2002) Mechatronics: from the 20th to 21st century. Control Eng Pract 10(8):877–886
9. Grimheden M, Hanson M (2005) Mechatronics—the evolution of an academic discipline in engineering education. Mechatronics 15(2):179–192
10. Schafer W, Wehrheim H (May 2007) The challenges of building advanced mechatronic systems. In: Future of software engineering (FOSE'07). IEEE, pp 72–84
11. Van Brussel H (Jan 1989) The mechatronics approach to motion control. In: Proceedings of the international conference of motion control: the mechatronics approach, pp 3–15
12. Shetty D, Kolk RA (2010) Mechatronics system design. CL-Engineering
13. Hewitt JR (Sep 1993) Mechatronics—the contributions of advanced control. In: Proceedings of the 2nd conference mechatronics and robotics
14. Dixit US, Hazarika M, Davim JP, Dixit US, Hazarika M, Davim JP (2017) History of mechatronics. A Brief Hist Mech Eng, 147–164
15. Grimheden M, Hanson M (2001) What is mechatronics?: proposing a didactical approach to mechatronics. In: 1st Baltic Sea workshop on education in mechatronics
16. Comerford R (1994) Mecha... what? [mechatronics]. IEEE Spectr 31(8):46–49
17. Isermann R (2005) Integrated mechanical electronic systems. Mechatron Syst: Fundam, 1–32
18. Bradley D (2018) Mechatronics: electronics in products and processes. Routledge

19. Harashima F, Tomizuka M, Fukuda T (1996) Mechatronics-"what is it, why, and how?" an editorial. IEEE/ASME Trans Mechatron 1(1):1–4
20. Venuvinod PK, Rao KP (Sep 1994) A mechatronic engineering degree course to meet the needs of Hong Kong. In: Proceedings of the international conference on mechatronics and machine vision in practice, pp 52–57
21. Acar M, Makra J, Penney E (eds) (1994) Mechatronics–the basis for new industrial development. Computational Mechanics
22. Sagi O, Rokach L (2021) Approximating XGBoost with an interpretable decision tree. Inf Sci 572:522–542
23. Mitchell TM (1997) Machine learning. McGraw-hill
24. Farhan A, Wei CL, Ahmed MT (2018) A qualitative overview of fuzzy logic in ECG arrhythmia classification. Int J Eng Work 5(11):232–239
25. Mosheiov G (2001) Scheduling problems with a learning effect. Eur J Oper Res 132(3):687–693
26. Gurel S, Selim Akturk M (2008) Scheduling preventive maintenance on a single CNC machine. Int J Prod Res 46(24):6797–6821
27. Dittrich MA, Uhlich F, Denkena B (2019) Self-optimizing tool path generation for 5-axis machining processes. CIRP J Manuf Sci Technol 24:49–54
28. Stone P et al (2016) One hundred year study on artificial intelligence: report of the 2015–2016 study panel, Recuperado el. 15
29. Lighthill J (1973) Artificial intelligence: a general survey. Science Research Council. Science Research Council (SRC), Government Report
30. Roque AS, Krebs VW, Figueiro IC, Jazdi N (2022) An analysis of machine learning algorithms in rotating machines maintenance. IFAC-PapersOnLine 55(2):252–257
31. Shaikat AS, Hussein MR, Tasnim R, Farhan A, Khan AMS, Mokhtar AH, Rahman MM (2023) Computer vision based automated attendance system using face recognition. In: Proceedings of the 6th industrial engineering and operations management Bangladesh conference Dhaka, Bangladesh
32. Abdulkareem NM, Abdulazeez AM (2021) Science and business. Int J 5(2):128–142
33. Kang Y, Cai Z, Tan CW, Huang Q, Liu H (2020) Natural language processing (NLP) in management research: a literature review. J Manag Anal 7(2):139–172
34. Chadebecq F, Vasconcelos F, Mazomenos E, Stoyanov D (2020) Computer vision in the surgical operating room. Visc Med 36(6):456–462
35. Nnodim TC, Arowolo MO, Agboola BD, Ogundokun RO, Abiodun MK (2021) Future trends in mechatronics. IAES Int J Robot Autom 1(10):24
36. Onge PS, Temme LA, McAtee A, O'Brien KJ, Byrd BK (2019) Evaluation of the commercial, off-the-shelf (COTS) King-Devick Eye Tracking System. Mil Med 184(Supplement_1):571–578
37. Hajduk Z (2017) High accuracy FPGA activation function implementation for neural networks. Neurocomputing 247:59–61
38. Kuru K, Yetgin H (2019) Transformation to advanced mechatronics systems within new industrial revolution: a novel framework in automation of everything (AoE). IEEE Access 7:41395–41415
39. Sun TH, Jia Z, Ge J, Sha R, Xie Y, Gu Y (2019) The multifunctional safety helmet for coal miners of the optomechatronics. Eur Int J Sci Technol 8(11):64–77
40. Lal A, Ardanuc S, Hoople JT, Kuo JC (2019) U.S. Patent No. 10,217,045. U.S. Patent and Trademark Office, Washington, DC
41. Maity R, Maity NP, Baishya S (2019) An efficient model of nanoelectromechanical systems based ultrasonic sensor with fringing field effects. IEEE Sens J 20(4):1746–1753
42. Zhang H, Marty F, Xia X, Zi Y, Bourouina T, Galayko D, Basset P (2020) Employing a MEMS plasma switch for conditioning high-voltage kinetic energy harvesters. Nat Commun 11(1):3221
43. Arkadii BS, Victor Z (2019) Scientific and methodological aspects in mechatronics. Global Energy 107(1):134–142

44. Negrete JC (2020) Nanotechnology an option in Mexican agriculture. J Biotechnol & Bioinform Res 2(2):1–3
45. Charfi A, Bouhamed SA, Bossé É, Kallel IK, Bouchaala W, Solaiman B, Derbel N (2020) Possibilistic similarity measures for data science and machine learning applications. IEEE Access 8:49198–49211
46. Chukwulozie OP, Chinagorom NE, Chukwuneke CJ, Nnodim CT (2017) The analysis of cigarate production using double exponential smoothing model. Acad J Sci 7(2):293–308
47. Alamleh H, AlQahtani AAS (2020) A weighting system to build physical layer measurements maps by crowdsourcing data from smartphones. IAES Int J Robot Autom 9(3):211
48. Guo W (2020) Explainable artificial intelligence for 6G: improving trust between human and machine. IEEE Commun Mag 58(6):39–45
49. Zhu G, Liu D, Du Y, You C, Zhang J, Huang K (2020) Toward an intelligent edge: wireless communication meets machine learning. IEEE Commun Mag 58(1):19–25
50. Wang X, Yang LT, Wang Y, Ren L, Deen MJ (2020) ADTT: a highly efficient distributed tensor-train decomposition method for IIoT big data. IEEE Trans Industr Inf 17(3):1573–1582
51. Nnodim CT, El-Bab AMRF, Ikua BW, Sila DN (2019) Estimation of the modulus of elasticity of mango for fruit sorting. Int J Mech Mechatron Eng 19(2):1–10
52. Nosakhale OS, Mwaniki C, Akorede MF (2019) Optimal sizing and analysis of a hybrid energy system for a community microgrid in Nigeria. J Eng Appl Sci 14(23):8769–8778
53. Salawu MD, Arowolo MO, Abdulsalam SO, Isiaka RM, Jimada-Ojuolape B, Olumide ML, Gbolagade KA (2020) A chi-square-SVM based pedagogical rule extraction method for microarray data analysis. Int J Adv Appl Sci 9(2):93–100
54. Olaniyi AS, Olaolu AM, Jimada-Ojuolape B, Kayode SY (2020) Customer churn prediction in banking industry using K-means and support vector machine algorithms. Int J Multidiscip Sci Adv Technol 1(1):48–54
55. Arowolo MO, Adebiyi MO, Adebiyi AA, Okesola OJ (2020) A hybrid heuristic dimensionality reduction methods for classifying malaria vector gene expression data. IEEE Access 8:182422–182430
56. Onokwai AO, Okonkwo UC, Osueke CO, Olayanju TMA, Ezugwu CA, Diarah RS, Nnodim TC et al (Dec 2019) Thermal analysis of solar box cooker in Omu-Aran Metropolis. In: Journal of physics: conference series, vol 1378, no 3. IOP Publishing, p 032065
57. Ajani AA, Oduol VK, Adeyemo ZK (2021) GPON and V-band mmWave in green backhaul solution for 5G ultra-dense network. Int J Electr Comput Eng 11(1):390
58. Nnodim CT, Fath El-Bab AM, Ikua BW, Sila DN (2021) Design, simulation, and experimental testing of a tactile sensor for fruit ripeness detection. In: Transactions on engineering technologies: world congress on engineering and computer science 2019. Springer Singapore, pp 59–73
59. Abdulsalam SO, Mohammed AA, Ajao JF, Babatunde RS, Ogundokun RO, Nnodim CT, Arowolo MO (2020) Performance evaluation of ANOVA and RFE algorithms for classifying microarray dataset using SVM. In: Information systems: 17th European, Mediterranean, and middle eastern conference, EMCIS 2020, Dubai, United Arab Emirates, November 25–26, 2020, proceedings 17. Springer International Publishing, pp 480–492

Sensors and Actuators

Nusrat Zahan Ramesha, Sadatuzzama Saagoto, Md Zonayed, Sayma Sultana Jhara, Rumana Tasnim, and Enamul Huq

Abstract Mechatronics engineering is a creative and dynamic field that focuses on cutting-edge technology for a variety of applications, which include high-speed manufacturing systems used in many contemporary companies. Modern manufacturing incorporates sophisticated technology, cutting-edge software, tools, and goods that improve quality of life by utilizing intelligent sensors and parameter controllers. In smart autonomous manufacturing, mechatronics has improved efficiency and product quality while decreasing technical faults. This chapter provides a detailed overview of recent mechatronics research with applications in manufacturing, agriculture, industrial automation, robotics, biomedical and assistive technology, human–machine interface, unmanned vehicles, energy, aerospace, and transportation. Production has significantly increased as a result of Industry 4.0, digitization, and artificial intelligence integration in mechatronics have significantly increased production. A significant factor in the change in the automobile industry has been the development of mechatronics in electric and driverless vehicles. Numerous research and development initiatives have the potential to advance the use of mechatronics, in line with forthcoming findings. Mechatronics-based technology encompasses nearly all aspects of intelligent industrial techniques and contributes to a high-quality lifestyle for almost everyone.

Keywords Sensor · Actuator · Digitization · Biomedical · Manufacturing

1 Introduction

Sensors and actuators are integral parts of a mechatronics system. Mechatronics has advanced in every aspect of contemporary or modern technology, from massive industrial systems to individual parts used in consumer products. Two of the most

N. Z. Ramesha (✉) · S. Saagoto · M. Zonayed · S. S. Jhara · R. Tasnim · E. Huq
Department of MTE, World University of Bangladesh, Dhaka, Bangladesh
e-mail: nusrat.zahan@mte.wub.edu.bd

M. M. Rahman et al. (eds.), *Mechatronics*, Emerging Trends in Mechatronics,
https://doi.org/10.1007/978-981-97-7117-2_2

important parts of mechatronic systems are electromagnetic sensors and electromechanical controllers. Sensors as well as actuators in mechatronics design and applications provide a practical understanding of the characteristics and operations of various electromagnetic and electromechanical devices. Mechatronics or engineering systems use them to sense undetermined signals and perceive unknown parameters of their environment. In essence, sensors are required to gather data from the system and monitor it. This will support in controlling and operating the system. Various control equipment needs the use of control actuators. Some devices use both; for example, MEMS uses microminiature sensors and actuators. This chapter presents an introduction to sensors and actuators. This chapter delves into the functions and uses of several typical sensors and actuators in mechatronic systems.

2 Classification of Sensors

Figure 1 shows the classification of sensors. Generally, two categories of sensors are available: analog and digital. When necessary, ADCs can convert the analog output from analog sensors into a digital signal. Some analog sensors are potentiometers, DC tachometers, capacitive sensors, piezoelectric sensors, strain gauges, torque sensors, etc. Digital sensors produce digital output as their name suggests. Some digital sensors are shaft encoders, light sensors, Hall effect sensors, MEMS sensors, etc.

2.1 *Analog Sensors*

An analog sensor measures continuous physical parameters, data, or quantities and converts them into electrical signals. Various mechatronics systems utilize the working principles and applications of certain analog sensors.

2.1.1 Potentiometer

The potentiometer, also known as the angular potentiometer or the linear potentiometer, is a crucial sensor that measures angular or linear displacement. Since the potentiometer draws no current, it is useful for monitoring currents. By balancing the unknown EMF against known potential differences, a potentiometer circuit compares and measures potential differences [1]. Ambrose Fleming, the vacuum diode's inventor, used this approach to measure currents up to 500 A in 1885 [2–4]. Xiao Dong used a potentiometer in the vehicle steering system [5]. This steering system includes sensors that transmit signals to inform the car of the driver's desired wheel movements. It is likely for the motors in the steering system to give the driver feedback on the car's condition. This system provides a safe and friendly environment. In reality, the industry uses a digital potentiometer as a light-independent

Fig. 1 Sensor classification diagram

resistor. An LDR (light-dependent resistor) is an alternative to the thermistor since it is less susceptible to changes in temperature and humidity [6]. Figure 2 displays a schematic circuit diagram of a potentiometer.

In the schematic below,

R1 and R2 = Resistors of the pot.

V = voltage difference between the input and output pin.

RL = the resistance of the load (a motor or an LED), when a pot is turned, the values of the resistors are changed too. Moving the pot one way will reduce R1 and increase R2. This outcome in different voltage values on the wiper (the one that goes to RL in the schematics below).

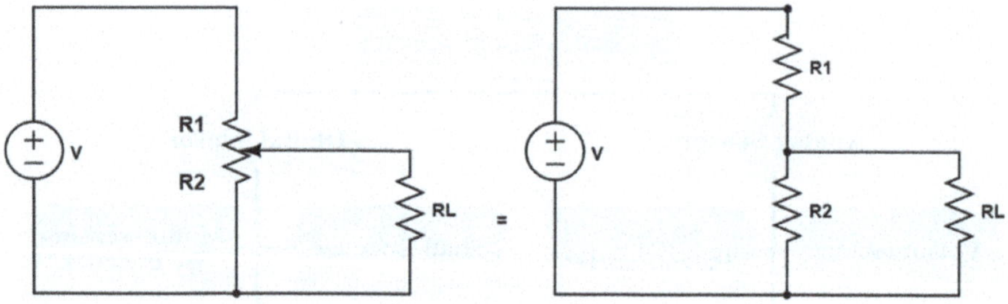

Fig. 2 A potentiometer schematic circuit diagram

$$V_L = \frac{R_2}{R_1 + R_2} \cdot V_s$$

where VL = voltage difference between the wiper and the output pin.

Potentiometers can control electrical signals and regulate resistance, commonly used in audio equipment like mixers, amplifiers, and guitars to regulate tone and volume [7].

2.1.2 Rotary Potentiometer

A rotary potentiometer, as shown in Fig. 3, adjusts the supply voltage to a specific portion of an electronic circuit. This potentiometer can adjust its resistance to electric current by rotating the knob or sliding the lever [8]. To determine the overall resistance, the rotary potentiometer consists of a resistive strip linked to terminals and a sliding contact connected to another terminal. By changing two resistances between three terminals 1, 2, and 3, the potentiometer acts as a regulator over these amounts or occurrences. This is done by the user moving its wiper. As shown in Fig. 3a, the maximum total resistance (RT) of an ideal potentiometer law is the sum of the resistances between terminals 1 and 3. Each terminal and wiper are regarded as ideal conductors, implying that they possess zero resistance [9].

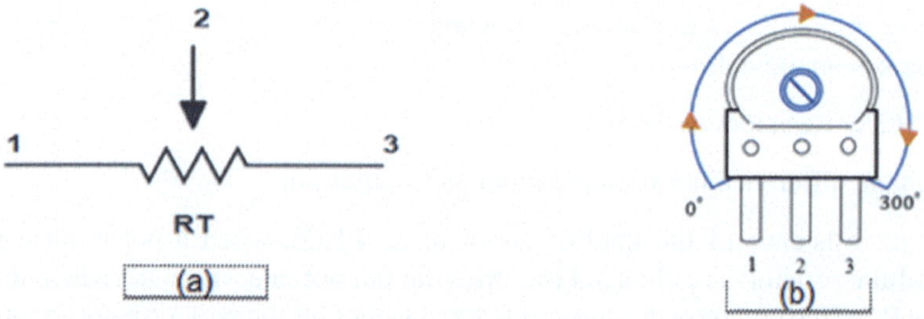

Fig. 3 **a** Potentiometer schematic symbol **b** Rotary potentiometer diagram with labeled terminals and the rotating shaft highlighted and range of rotation indicated, the shaft shown at 0°

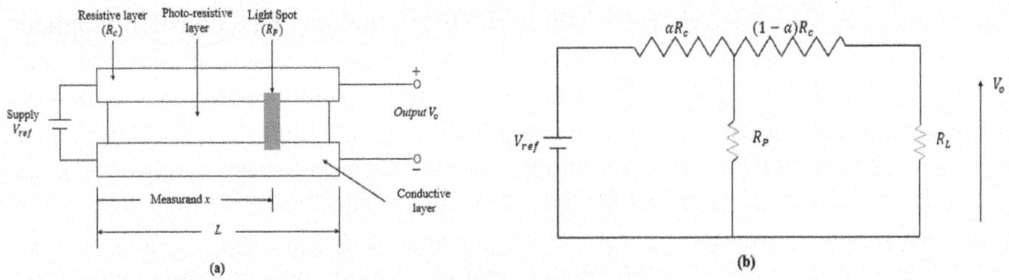

Fig. 4 **a** An optical potentiometer **b** Equivalent circuit (α = χ/L)

Many equipment functions use rotary potentiometers [10], including robotic hand control systems, stiffness readouts in musculoskeletal humanoid robots [11], volume control, and position sensors in motor control [12].

2.1.3 Optical Potentiometer

Optical potentiometers, or optical position sensors, are electronic devices that measure and control position or displacement [13]. An optical potentiometer uses a fluorescent optical fiber sensor to measure an object's position [14]. The potentiometer employed a fluorescent optical fiber with a track diameter of 1 mm. The optical cable was terminated by photodiodes with a p-i-n structure, while a light-emitting diode modulated at a frequency of 5 kHz was used as the excitation slider. According to Beer's law,

$$\mathrm{I(d)} = \mathrm{IOe} - \alpha \mathrm{d}$$

where I(d) is the intensity of light at any distance, d, from the detector, I is the constant multiplier intensity, and a is the attenuation constant [7]. The optical potentiometer, shown schematically in Fig. 4, is a displacement sensor. A layer of photo-resistive material sits between a regular resistive layer and a conductive layer [15]. In the following Figure, Rc = total resistance, Rp = Resistance of light projected area, RL = resistance of the load vref = Supply voltage of pot, and L = length of resistive layer.

Optical potentiometers are versatile devices that find applications in many industries and fields [16], especially for accurate measurement of displacements and velocities [17], as well as in robotics, medical fields, etc. [18].

2.1.4 DC Tachometer

A tachometer uses an analysis of the relative movement between the magnetic field and the shaft to determine the angular velocity, or rotational speed, of a machine. The velocity of the coils is precisely proportional to the speed of the shaft. This velocity induces an electromagnetic force (EMF) in the magnet's constant magnetic field. There are two varieties of tachometers on the market: electronic and mechanical. Compared to mechanical tachometers, the electrical tachometer provides more advantages by converting rotational speed into electrical voltage. The machine's shaft is connected to the DC tachometer generator, which measures the machine's speed. Some of the important parts of the DC tachometer generator are the moving coil voltmeter, brushes, armature, commutator, permanent magnet, and variable resistor. The DC tachometer operates by moving the conductor through a magnetic field, which induces an electromotive force (EMF). The conductor's magnetic flux and the shaft's velocity determine the magnitude of the induced electromotive force (emf). Brushes convert the force, proportional to shaft speed, into direct current. The voltmeter connects in series with the resistance to regulate the heavy armature current, while the induced voltage polarity controls the shaft motion. To measure speed, permanent-magnet transducers use electromagnetic induction between a permanent magnet and a conducting coil [15].

2.1.5 Capacitive Sensor

A capacitive sensor detects liquids or solids without physical contact. To detect these targets, the capacitive sensor's detecting end generates an electrical field. This sensor detects targets that interfere with the electrical field. Capacitors can detect paper, plastic, glass, cloth, and wood. A capacitor sensor detects oil, paint, water, etc. A capacitive sensor functions like a capacitor. This sensor connects a metal plate on the sensing face to an oscillator circuit, and the detected target becomes the capacitor's next plate. Capacitive sensors generate electrostatic fields, unlike inductive sensors. A capacitive displacement sensor is a type of capacitive sensor. Capacitive displacement sensors are contactless instruments that can accurately measure the position and movement of any conductive object with great precision. In addition, they have the capability to gauge the thickness or density of materials that do not conduct electricity [19]. Figure 5 shows two different architectures of capacitive sensing.

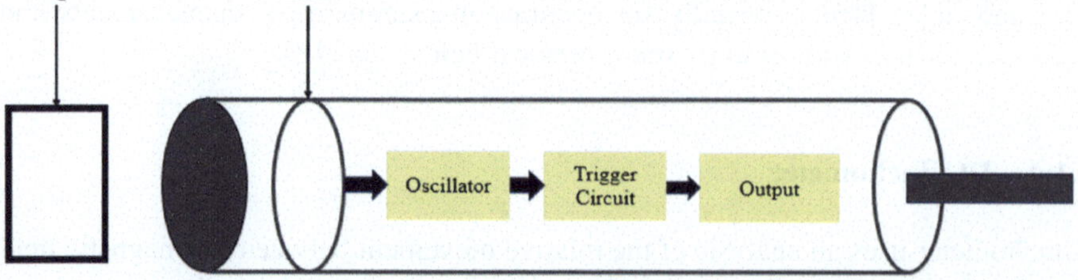

Fig. 5 Capacitive sensor

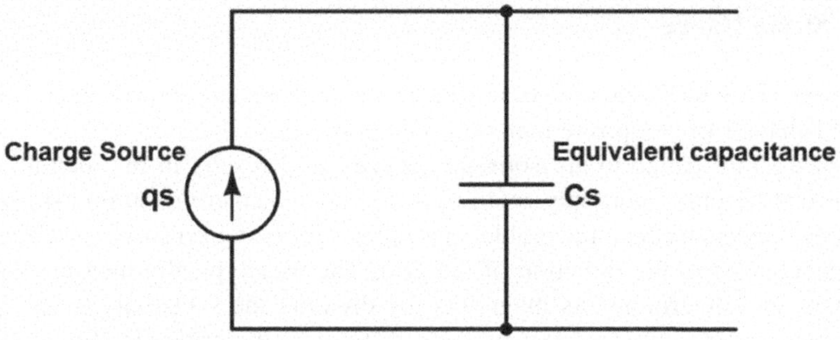

Fig. 6 Piezoelectric sensor

Some of the applications of capacitive sensors are motion sensing [20], gauging and metrology [21], object detection [22], liquid level sensing [23], material testing, environmental sensing [24], flow sensing [25–30], and biomedical applications [31–34].

2.1.6 Piezoelectric Sensors

A piezoelectric accelerometer is a device that detects and quantifies changes in mechanical variables such as vibration, acceleration, and mechanical shock. It achieves this by harnessing the piezoelectric effect of specific materials. The piezoelectric effects can be classified into four distinct categories: the longitudinal effect, shear effect, transverse effect, and hydrostatic effect. However, the fabrication of CNF films is characterized by their physical, dielectric, and electromechanical behavior. Industry compares the nonlinearity and hysteresis of piezoelectric sensors with those of polyvinylidene fluoride (PVDF), a reference piezoelectric polymer [35, 36]. Figure 6 presents a piezoelectric sensor, where qs represents the source's charge and Cs stands for capacitance.

Piezoelectric actuators framework a significant portion of the transducer market, with applications in automotive systems, active vibration damping, medical imaging, precision positioning, industrial machinery, and consumer electronics [36].

2.1.7 Accelerometer

An accelerometer measures an object's or body's acceleration in its immediate reference frame. An accelerometer doesn't measure acceleration in terms of coordinates. Accelerometer sensors have diverse applications in various electronic products, including smartphones and wearable gadgets. A micro-machined piezoelectric sensing element combines with a miniature wireless accelerometer to measure mechanical vibrations [37].

2.1.8 Strain Gauge

With the growing utilization of strain gauges, the techniques and principles controlling the behavior of composite materials are being adjusted accordingly. The application of a strain gauge to an isotropic material [38] results in the manifestation of a thermal output, commonly referred to as apparent thermal strain. Transverse sensitivity denotes to the unfavorable consequence of being sensitive to strains that are perpendicular to the direction of the grid. The user typically measures it as a percentage and determines its magnitude by dividing the sensitivity in the direction perpendicular to the grid axis by the sensitivity in the grid axis' direction. The foil strain gauges exhibit a transverse sensitivity that ranges from a few percent. Self-temperature compensated (S-T-C) strain gauges have zero thermal output on tested materials, causing thermal compensation loss. Even with a correctly chosen strain gauge, orthotropic behavior and transverse sensitivity combine to leave behind thermal output. Materials like composite behave orthotopically, which also produce the different transverse sensitivity effects [39, 40].

2.1.9 Torque Sensor

The unique wrist force or torque sensors used by the industries have four degrees of freedom (DOF) for haptic-based human–computer interface systems. The 4 DOF sensor is more cost-effective than conventional 6 DOF sensors because of its simplified structure and reduced strain gauge demands. Six-axis MEMS force torque sensor is used by industry to measure the torque components on various axes. Two force components and one torque component must be measured, to characterize magnetic material properties. Figure 7 presents a basic torque sensor, with fx, fy, and fz representing generalized forces or moments acting on the torque sensor in the x, y, and z axes, respectively. Torque sensors are commonly used in robotics to control joint movements. Various industries, including automotive, also use torque sensors [41–44].

2.1.10 Gyroscopic Sensors

The gyroscope device attaches a disk to a frame and detects angular velocity when the frame rotates. There are various classes of gyroscopes that depend on the physical and technological aspects of their operation. A gyroscope can work by itself or as part of a more complicated system, such as an Inertial Measurement Unit (IMU), gyrocompass, attitude heading reference device, or method for navigation. A gyroscope's operational mechanisms include rotational movement along specific axes. The gyroscope uses capacitance variations to detect the displacements of the object.

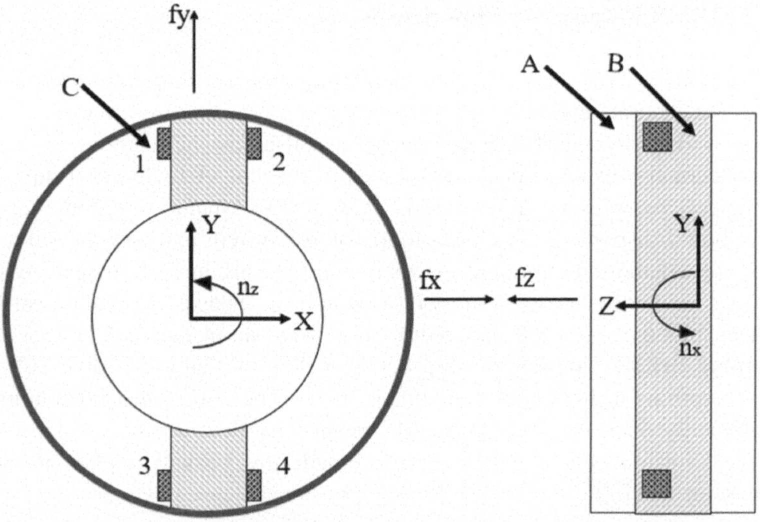

Fig. 7 Basic torque sensor structure **a** solid disks, **b** elastic element, and **c** strain gauge

2.1.11 Thermo-Fluid Sensors

There are different types of sensors that involve fluid flow, thermodynamics, heat transfer, and pressure sensors.

Fluid Flow: Sensors to detect variables such as pressure, velocity, and flow rate are used in fluid flow applications, which provide valuable information for efficient system operation, anomaly identification, and performance improvement [45]. These sensors ensure system efficiency, identify anomalies, and enhance overall system performance.

Thermodynamics: Sensors are used to measure temperature, pressure, and other factors to monitor and control industrial operations, refrigeration units, and heating and cooling systems [46].

Heat transfer: Heat-transfer-detecting sensors are essential in applications like HVAC systems, which monitor temperature differentials to maintain a suitable indoor environment [47]. Additionally, renewable energy fields like solar power plants use sensors to monitor and optimize the power output.

Pressure: It is essential to comprehend and analyze the pressure exerted on the wings of insect-like flying robots in order to quantify their pressure accurately. Utilize a pressure sensor with a weight that is at least one-tenth of the weight of the wings [48].

Thermo-fluid sensors are utilized in various sectors and situations where it is crucial to monitor and manage temperature and fluid flow [49].

2.1.12 MEMS Respiratory Flow Sensor

MEMS respiratory flow sensors play a crucial role in modern respiratory care because they provide accurate and real-time measurements. A sensor information feedback system uses a high-speed MEMS flow sensor to self-adjust the solenoid valve's open time to accurately dispense desired volumes of reagent. A high-speed liquid flow sensor detects the pressure differential across a flow channel for both static and dynamic liquid dispensing. A closed-loop control system assesses the valve open time for any fluctuations in liquid viscosity and pressure in each dispensing cycle. Techniques for monitoring respiratory flow include the measurement of transthoracic impedance, blood oxygen and carbon dioxide concentration, and direct measurement of breathing airflow. The relationship between flow rate and measured voltage in resistive breathing airflow measurements is linear. The cost of instrumentation can be significantly decreased with a MEMS respiratory flow sensor. A piezoelectric cantilever deform occurs when the exhaled breath flow velocity applies pressure to the sensor surface [50].

2.1.13 Thermocouple

When a variation in temperature causes an electrical potential difference that arises between two different metallic materials, thermocouples function by harnessing the Seebeck effect at that time. The voltage is directly proportional to the temperature difference between the measuring junction and the reference junction. Many industries use these materials due to their high-temperature resistance, durability, and suitability for high-temperature environments [51, 52], including manufacturing, automotive, and aerospace [53, 54]. Figure 8 shows a thermocouple.

2.1.14 Thermistor

A thermistor is a type of resistor that acts as a thermometer by changing its resistance based on temperature. Thermistor sensors provide a change of resistance, and the typical power range is 1 μW to 10 mW [55]. In reaction to variations in heat transfer, which for a given geometry is mostly a function of the air speed and temperature, thermistors can offer velocity data. Thermistors have more stability, resilience, and accuracy at low velocities than conventional thermal anemometers like hot wires [56]. The creation of thermos-resistors typically involves the use of semiconductor materials suitable for their temperature-dependent resistance change. These materials are typically doped ceramics or a mixture of several metal oxides, including cobalt, copper, magnesium, manganese, titanium, and nickel [57]. The temperature coefficient of resistance (NTC) of thermocouples is typically negative. A substance with a high degree of temperature sensitivity makes up a thermistor. The material that comprises a thermistor has a high degree of temperature sensitivity. The material's resistance is directly correlated with its temperature-calibrated expansion and

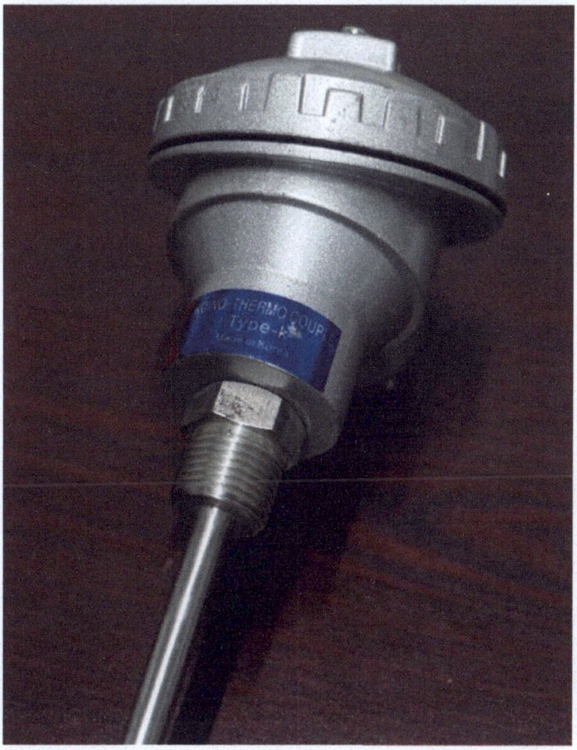

Fig. 8 Thermocouple

contraction. Figure 9 presents a schematic diagram of a thermistor. In this figure, the brightness of the light bulb varies with the measured temperature [58].

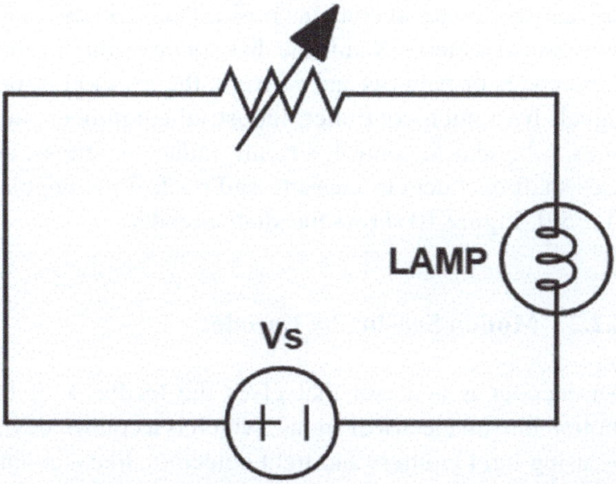

Fig. 9 Schematic diagram of thermistor

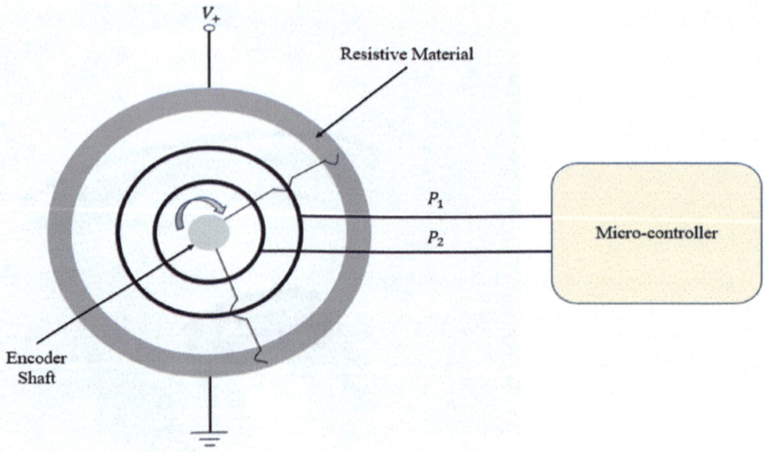

Fig. 10 Shaft encoder

2.2 Digital Sensors

Industries use different types of analog sensors for different mechatronics systems. This part of the book highlights the working principles and applications of analog sensors.

2.2.1 Shaft Encoder

An encoded disk which is a resistive material and a decoding device combine to form an absolute shaft encoder, also known as a rotary encoder. The encoded disk or the resistive material contains positional information obtained using linear codes or mechanical patterns. While the disk rotates with the object, a decoding unit interprets these codes or patterns and converts the location information into analog or digital signals by a microcontroller. Industrial equipment, such as machine tools, automobiles, robotic arm control, aircraft, radars, antennas, and conveyor systems, widely uses shaft encoders to measure and control the angular position of rotating objects [17, 59]. Figure 10 shows the shaft encoder.

2.2.2 Motion Sensing by Encoder

An encoder is a sensor that gives the feedback of the system. Encoders convert motion into an electrical signal, which is a control device in a motion control system by using light emitters and light detectors, like a counter or PLC, and can interpret. The encoder transmits a return signal to determine the position, count, speed, or

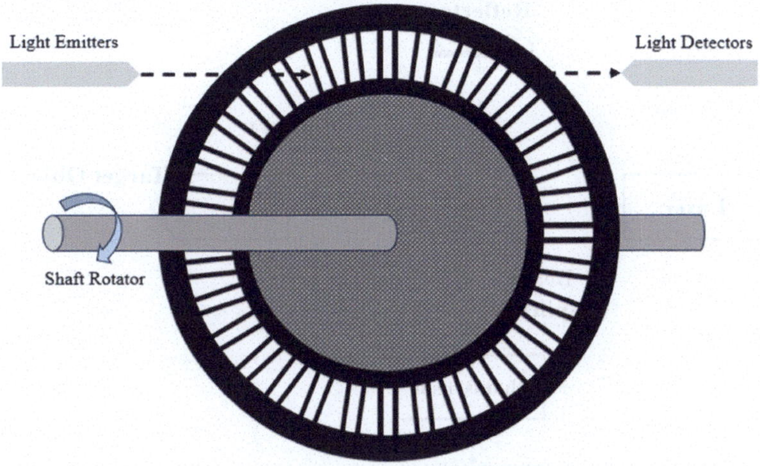

Fig. 11 Motion sensing by encoder

direction. Elevators, pointing mechanisms, and servomotors use this encoder [60, 61]. Figure 11 shows motion sensing by encoder.

2.2.3 Laser Doppler Interferometer

Laser Doppler Velocimetry (LDV) is a method that enables the precise measurement of velocity at a specific location within a flow field, with a focus on capturing rapid changes over time. To examine the material properties, calibrate sensors, and measure force in dynamic circumstances, the laser Doppler interferometer (LDI) can be utilized. It has a wide range of applications in various fields such as Flow Research, Automation, Medical Applications, Navigation, Calibration and Measurement Spectroscopy, Metrology, Gravitational Wave Detection, NVH Optimization (Noise, Vibration, Harshness), Measurement of Material Velocity, etc. [62–64]. Figure 12 shows Laser Doppler Interferometer.

2.2.4 Light Sensors

A light sensor generates an output signal that represents the brightness of light by measuring the radiant energy within a particular frequency range known as the "light" spectrum. This spectrum includes all of the frequencies from "infrared" to "visible" and even "ultraviolet" light. Figure 13 shows four types of light sensors that include photoresistor, photodiode, phototransistor, and photo-FET. Light sensors are used in consumer electronics, streetlights, automobiles, security systems, photography, solar panels, display backlight control, proximity sensors, agricultural monitoring systems, photoelectric rangefinders, and memory material illumination sensors [65].

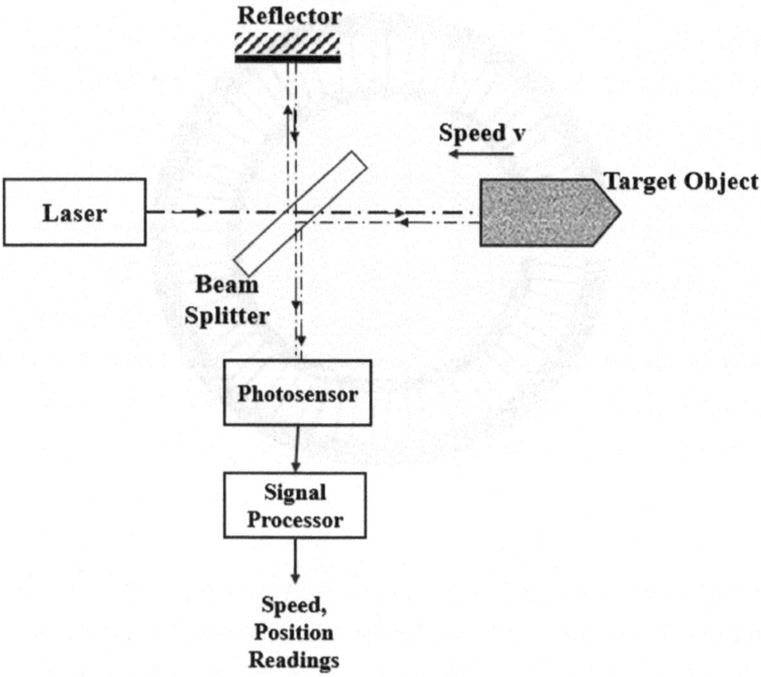

Fig. 12 Laser doppler interferometer

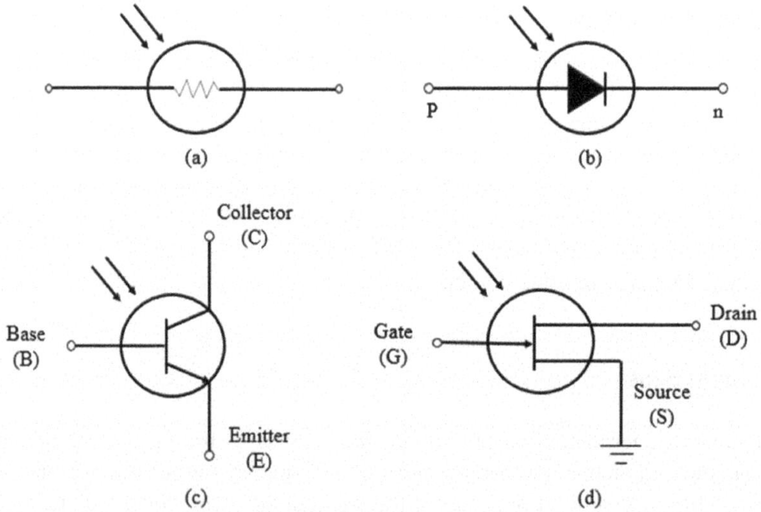

Fig. 13 Four types of light sensors

Photoresistor: At high light on their sensitive surfaces, photoresistors, which are passive electronic parts, lose resistance. They are also known as photocells, light-dependent resistors (LDRs), or photo-conductive cells. This phenomenon is known as photoconductivity. Researchers are using silica nanopillars as a substrate for cadmium sulfide (CdS) in photoresistors for the first time. The substrate with nanopillars, which has a high surface ratio, can enhance the amount of sensitive material and the light-responsive surface area. Consequently, this significantly enhances the photosensitivity of the CdS photoresistor [66]. Photoresistor sensors find use in various fields such as light measurement and detection systems, medical practice, optical communication, security monitoring, analytical instrumentation, industrial control, and home automation [67, 68].

Photoresistor: A photodiode is a type of semiconductor diode that can detect and respond to several forms of photon radiation, including visible light, infrared or ultraviolet radiation, X-rays, and gamma rays [69]. A photodiode resistor is also useful in weather forecasting applications because it can measure solar irradiation [70, 71]. Systems for light measurement and detection, environmental and security monitoring, analytical instrumentation, industrial control, home automation, and modern agriculture also utilize it.

Phototransistor: A phototransistor is a device that converts incoming photons into electrons at the base of a bipolar transistor. The flow of current from the base causes a larger current to flow between the collector and emitter just like the other transistor. A circuit then detects the increase in current. Electronic devices and industrial automation systems use phototransistors as ambient light sensors to detect objects, measure distances, and control processes based on light signals [72].

Photo-FET: A photo-FET, or photo-field-effect transistor, is a specialized semiconductor device that modulates its conductivity in response to light. It works by adjusting the conductivity of the semiconductor channel in response to light, where photodiodes and phototransistors both use the photoelectric effect. The gate-source voltage is adjusted to create this modulation [73]. Photo-FETs can be used in optical communications and high-speed light sensing applications.

2.2.5 Hall Effect Sensor

A Hall effect sensor, also known as a Hall sensor or Hall probe, incorporates one or more Hall elements. These elements utilize the Hall effect, named after scientist Edwin Hall, to generate a voltage that is directly proportional to a specific component of the magnetic field vector B. A low voltage output, usually in the range of a few microvolts per gauss, is generated by the Hall Effect magnetic sensors. As a result, these sensors often come with built-in high-gain amplifiers [74]. There are two classifications of Hall Effect sensors: one that produces analog output and another that generates digital output. The analog sensor comprises a voltage regulator, a Hall element, and an amplifier. An analysis of the circuit diagrams reveals that the

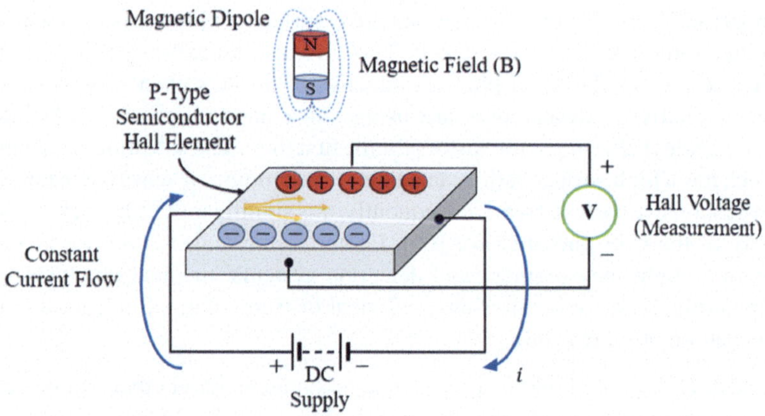

Fig. 14 Hall effect sensor

sensor's output is analog and has a direct correlation with either the Hall Element output or the intensity of the magnetic field. These sensors are suitable and used for monitoring proximity because of their consistent linear output. The Hall effect sensors have advantageous characteristics like thermal stability and high sensitivity. Moreover, they have very low detection limits. Because of their uniqueness, planar Hall effect sensors are used in nano-Tesla (nT) magnetometers, current sensing, or low magnetic moment detection in lab-on-a-chip devices [75, 76]. Figure 14 shows Hall effect sensor.

2.2.6 Ultrasonic Sensors

An ultrasonic sensor measures the distance between itself and an object by using ultrasonic sound waves that use a transducer to produce and detect ultrasonic pulses that identify how far away the object is. Sound waves are produced and emitted by a transmitter using piezoelectric crystals, and a receiver captures any reflections of these waves. The sensor captures and transforms the sound that bounces off the intended object into digital or electronic signals for later analysis or manipulation. Assessing surface characteristics, identifying object position, and computing object velocity, ultrasonic sensors have found extensive applications in such tasks. Also, in-vehicle detection uses an ultrasonic sensor [77]. The observer can use the ultrasonic sensors to determine the position and velocity of moving objects like cars, buses, etc. The vehicle passes the ultrasonic sensor at the roadside [78]. Some applications of ultrasonic sensors are industrial sensing, precision actuation, material processing and manufacturing, medical imaging, and medical therapy [79–81]. Figure 15 shows the ultrasonic sensor.

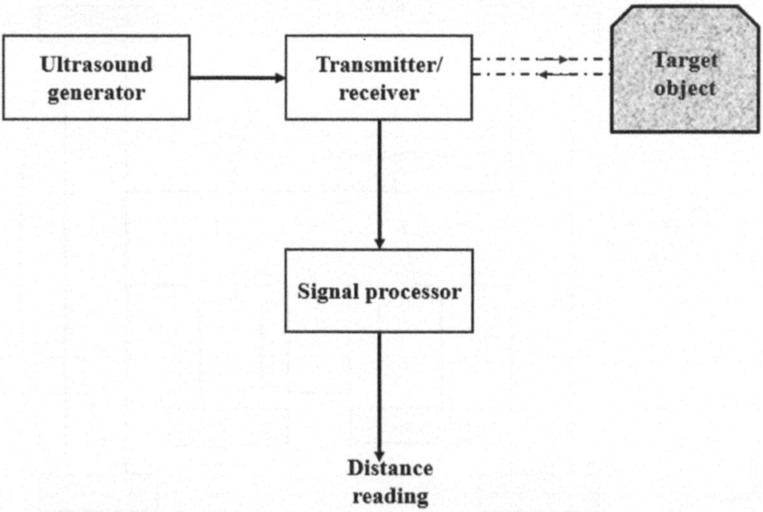

Fig. 15 Schematic diagram of ultrasonic sensors

2.2.7 MEMS Sensors

MEMS sensors are compact devices utilized for the measurement of physical parameters such as pressure, temperature, acceleration, and magnetic fields. Advantages like compactness, light weightiness, superior performance, convenient mass production, and affordability are offered by MEMS technology [82]. Depositing thin material layers onto a silicon substrate and then selectively removing areas to create 3D formations like diaphragms, beams, levers, springs, and gears are involved in MEMS IC manufacturing. Tilting the sensor causes a change in electric potential, quantified as capacitance. The industry can modify the aforementioned signal to generate a consistent output signal in either digital, 4–20 mA, or VDC format.

There are four types of MEMS sensors:

- MEMS accelerometer.
- MEMS gyroscopes.
- MEMS pressure sensors.
- MEMS magnetic field sensors.

Using a MEMS capacitive force sensor, researchers investigated the mechanical properties of soft hydrogel microcapsules as a protein delivery mechanism. Many industries like petrochemical, industrial, and construction services sectors, power generation, aerospace, defense, and healthcare currently utilize MEMS. They are also essential parts of computers, phones, digital cameras, game consoles, and automobiles, and they play a significant role in daily life [83, 84]. Figure 16 shows MEMS sensor.

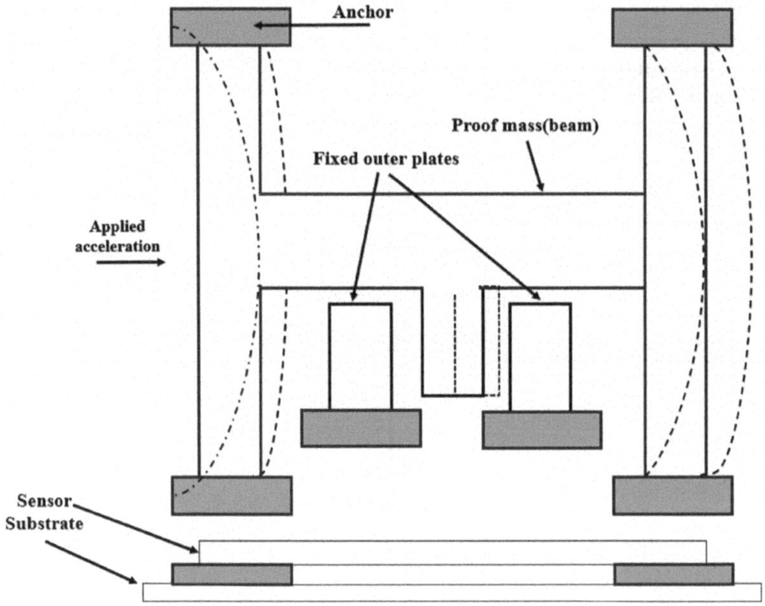

Fig. 16 MEMS sensors

2.3 Sensor Fusion

Sensor fusion is the combination of sensor data or data generated from other sources in order to generate information with reduced uncertainty compared to using these sources individually. Different types of techniques, such as the extended Kalman filter, combine data from the Global Positioning System (GPS) and the inertial navigation system (INS). This is particularly useful, for example, in determining the orientation of an airplane with inexpensive sensors. This is one of the uses of sensor fusion [85]. Railway systems also incorporate sensor fusion [86]. Automation, automotive driver assistance systems, autonomous robots, C3I, etc. are some other applications of sensor fusion [87].

3 Actuator

An actuator is a machine component that, when provided with an electrical, pneumatic, or hydraulic input in a system (referred to as an actuating system), creates force, torque, or displacement, usually in a controlled manner. An actuator transforms such an input signal into the necessary kind of mechanical energy [88].

3.1 Stepper Motors

A stepper motor is an electric motor that rotates in a sequence of small angular steps rather than continuously. It is also known as stepping motor. The stepper motor is an electromechanical device that transforms electrical energy into mechanical energy. It consists of a permanent-magnet rotor and an electromagnet stator and it follows the principle of electromagnetism. It can generate a magnetic field by supplying power to the stator coil that in turn causes the rotor to rotate in response to the rotating magnetic field. This motor operates on a fundamental functioning principle. Stepper motors play a crucial role in the field of robotics. Robotic applications necessitate accurate and exact motions. The stepper motor drive specifically builds and rigorously tests its software and hardware circuits to accommodate various types of stepper motor movements [89, 90].

3.2 DC Motor

A direct current motor or DC motor is an electrical device that uses a direct current to create a magnetic field in order to convert electrical energy into mechanical energy. The stator of a motor is usually placed outside and the rotor is outside.

The stator contains electromagnetic windings or permanent magnets, while the rotor's coil windings are powered by DC current. Rotor magnets are attracted to and repelled by the magnetic field produced in the stator by DC current. After that rotor spins and motor employ commutators to turn the rotor. The rotor stops spinning when it aligns with the magnetic field, but the magnetic field is reversed since the commutator reverses the stator current. This keeps the rotor rotating. The right image shows a schematic of the DC motor. A study goes into great detail about a five-phase brushless DC (BLDC) motor that was made for an electro hydrostatic actuation system (EHA) that works well with thin, well-designed wings [91]. Figure 17 shows a DC motor. The main applications of DC motors include the manufacture of pulp, paper, and paperboard; the propulsion of electric vehicles; textile industries; and public transportation like subway and trolley systems [92, 93].

3.3 Synchronous Motor

A synchronous electric motor precisely synchronizes its shaft's rotation with the frequency of the provided electrical current in a stable condition. Due to their ability to maintain speed, coordinate with the grid, and support reactive power, synchronous generators are widely used in Power Plants for electricity generation. The time it takes for one complete shaft rotation is equivalent to a full number of AC cycles [94]. High-performance driving applications such as industrial robots and machine

Fig. 17 DC motor

tools frequently utilize synchronous motors due to their high power density, high torque-to-inertia ratio, and low maintenance requirements [95, 96].

3.4 Hydraulic Pumps and Motors

A hydraulic pump is a mechanical device that converts mechanical power into hydraulic energy. In the form of flow and pressure, it specifically converts the hydrostatic energy. Agricultural equipment uses hydraulic pumps and motors to enhance its efficiency. The most significant application of hydraulic pumps and motors is to improve the overall efficiency of agricultural equipment, such as tractors, harvesters, planters, fertilizers, sprayers, and attachments [97, 98]. Figure 18 shows hydraulic pump.

3.5 Hydraulic Valves

A hydraulic valve is a mechanical device that controls the movement of hydraulic fluid inside a hydraulic system. Their high-pressure hydraulic systems are commonly characterized as ranging from 200 bar to an average of 700 bar or higher. The system uses the hydraulic valves to regulate the fluid's flow rate. The system also monitors and controls the fluid pressure and direction. Electro-hydraulic control valves are critical hydraulic components used in industrial and aerospace applications to regulate electro-hydraulic motion. Electro-hydraulic control valves are becoming increasingly digital, integrated, and intelligent to meet the demands of Industry 4.0, due to

Fig. 18 Hydraulic pump

the advancements in automation, digital technology, and communication technology [99].

4 Sensor and Actuator Integration in Mechatronic System

The integration of sensors and actuators into mechatronic systems requires a series of crucial stages.

4.1 Signal Acquisition and Conditioning

Sensors perceive physical characteristics and help transform them into electrical impulses. These signals frequently require conditioning, such as amplification or filtering, in order to be functional. Signal conditioning includes tasks like boosting weak signals, getting rid of noise, and adjusting signals to a range that works with the analog-to-digital converter (ADC) if the system uses digital processing [100].

4.2 Signal Processing

Microcontrollers and microprocessors handle the processed signals received from sensors. They implement control algorithms to ascertain the suitable reaction. Control algorithms can range from basic, such as proportional control, to more intricate ones like PID control and model predictive control [101].

4.3 Actuation Control

The processor generates signals and transmits them to actuators in order to produce the intended physical reaction. Power electronics, such as H-bridges for motor control and valve drivers for hydraulic actuators, regulate power flow to actuators according to control signals [102].

4.4 Feedback and Closed-Loop Control

For achieving an accurate control, Feedback loops are crucial in mechatronics system. The user compares the required values by constantly monitoring the sensor data. Error correction involves the manipulation of actuator signals to minimize the discrepancy between the measured and desired states. The integration process utilizes both wired and wireless communication protocols [103]. Common wired communication protocols for exchanging data between sensors, actuators, and controllers include I2C, SPI, UART, TFT, and CAN buses. When establishing physical connections is not feasible, wireless communication applications use Bluetooth, Zigbee, and Wi-Fi [104]. Software integration employs embedded software such as real-time operating systems or firmware to handle tasks such as sensor data collection, processing, and control logic implementation, while MATLAB/Simulink facilitates the simulation and modeling of control methods [105]. Here is an illustration of a system that integrates both a sensor and an actuator. An ADC process detected the finger presence from a capacitive touch sensor and servo motor into a digital signal. A microprocessor determines if the touch threshold has been surpassed or not by interpreting the digital signal by enhancing the device's functionality [106]. To adjust the position of a servo motor [107], the microcontroller uses a control algorithm and regulates its movement through PWM signals. The motor installs a position sensor for precise control, enabling closed-loop control via feedback from the microcontroller. This ensures precise and efficient motor operation.

5 Conclusion

The integration of sensors and actuators in designing a mechatronics system is essential for the advancement of many industries, including robotics, automation, and manufacturing. This integration enhances the system's functionality and performance. Moreover, mechatronics' multidisciplinary character offers a wide choice of options for completing particular functionality. To attaining optimal system performance in mechatronics system, integrating sensors and actuators is essential. Sensors can sense unknown characteristics of a mechatronics or engineering system and its surroundings. They are able to monitor the system and collect data about it. This will support system operation and control. Actuators, such as stepper motors and hydraulic pumps, are necessary for plants to function. Control actuators are required for various control equipment to operate. Furthermore, mechatronics, as an integrated technology, advances due to the ongoing development of microelectronics and other technologies. In today's engineering curriculum, mechatronic system design and comprehension are becoming more and more crucial to stay up with the rapidly changing technological landscape.

References

1. Enderle J, Bronzino J (2011) Introduction to biomedical engineering. Academic Press
2. Mitchell J, Griffiths H, Boyd I (2006) Sir John Ambrose Fleming—his involvement in the development of wireless. https://doi.org/10.1002/0471783021.ch10
3. Britannica T editors of Encyclopaedia (14 Apr 2024) Sir John Ambrose Fleming. Encyclopedia Britannica
4. John Ambrose Fleming (n.d.) John Ambrose Fleming. Magnet Academy
5. Wu X, Zhang M, Xu M (2019) Active tracking control for steer-by-wire system with disturbance observer. IEEE Trans Veh Technol 68(6):5483–5493
6. Banihani SM, Dutkin M, Ali S, Arikatla VS, Sankaranarayanan G, De S (2009) Plug-and-Play tool handles for laparoscopic surgery simulators. Stud Health Technol Inform 142:28–30
7. Lang SR, Ryan DJ, Bobis JP (1992) Position sensing using an optical potentiometer. IEEE Trans Instrum Meas 41(6):902–905
8. Aguirre-Manríquez IV, García-Guzmán MÁ, Razón-González JP, Cano-Lara M (2022) Basic operation of rotary potentiometer A, B, C and W. J Innov Eng 6(19):1–5
9. Holmes B, Van Walstijn M (Sep 2019) Potentiometer law modelling and identification for application in physics-based virtual analogue circuits. In: Proceedings of the 22nd international conference digital audio effects (DAFx-19)
10. Wood B, Hussett D, Cattaldo O, Rohr T (2005) SADM potentiometer anomaly investigation
11. Masahiko O, Nobuyuki I, Yuto N, Masayuki I (2010) Stiffness readout in musculo-skeletal humanoid robot by using rotary potentiometer. In: SENSORS, 2010 IEEE, Waikoloa, HI, USA, pp 2329–2333. https://doi.org/10.1109/ICSENS.2010.5690907
12. Kanchi RR, Gosala N (2010) Design and development of an embedded system for testing the potentiometer linearity. Sens Transducers 112(1):107
13. Kwon Y, Kim W (2014) A low-cost rotary optical potentiometer mechanism based on opto-electronics technique. In: Proceedings of the ASME 2014 dynamic systems and control conference. Volume 3: industrial applications; modeling for oil and gas, control and validation, estimation, and control of automotive systems; multi-agent and networked systems; control system design; physical human-robot interaction; rehabilitation robotics; sensing and actuation for control; biomedical systems; time delay systems and stability; unmanned ground and surface robotics; vehicle motion controls; vibration analysis and isolation; vibration and control for energy harvesting; wind energy. San Antonio, Texas, USA, October 22–24, 2014. V003T44A003. ASME
14. Laguesse M (1989) Optical potentiometer using fluorescent optical fiber for position measurement. Appl Opt 28:5144–5148
15. de Silva CW (2003) Sensors for control. In: Meyers RA (ed) Encyclopedia of physical science and technology, 3rd ed. Academic Press, pp 609–650. ISBN 9780122274107
16. Koutilellis GD, Economou A, Efstathiou CE (2016) A potentiostat featuring an integrator transimpedance amplifier for the measurement of very low currents—proof-of-principle application in microfluidic separations and voltammetry. Rev Sci Instrum 87:3

17. Paredes F, Herrojo C, Martín F (2021) Position sensors for industrial applications based on electromagnetic encoders. Sensors 21(8):2738
18. Osman D, Li W, Du X, Minton T, Noh Y (2022) Miniature optical joint measurement sensor for robotic application using curvature based reflecting surfaces. IEEE Sens Lett
19. Ortmanns M, Buhmann A, Manoli Y, 1.16—Interface Circuits, Editor(s): Gianchandani Y, Tabata O, Zappe H (2008) Comprehensive microsystems. Elsevier, pp 465–493. ISBN 9780444521903
20. Arshad A, Khan S, Alam AZ, Kadir KA, Tasnim R, Ismail AF (May 2017). A capacitive proximity sensing scheme for human motion detection. In: 2017 IEEE international instrumentation and measurement technology conference (I2MTC). IEEE, pp 1–5
21. Acero R, Santolaria J, Brau A, Pueo M (2016) Virtual distances methodology as verification technique for AACMMs with a capacitive sensor based indexed metrology platform. Sensors 16(11):1940
22. Ćika D, Martinović T, Džapo H (May 2015) Measurement system for object detection based on multielectrode capacitive sensor. In: 2015 38th international convention on information and communication technology, electronics and microelectronics (MIPRO). IEEE, pp 137–142
23. Terzic E, Nagarajah CR, Alamgir M (2010) Capacitive sensor-based fluid level measurement in a dynamic environment using neural network. Eng Appl Artif Intell 23(4):614–619
24. Shemelya C, Cedillos F, Aguilera E, Maestas E, Ramos J, Espalin D, MacDonald E et al (Nov 2013) 3D printed capacitive sensors. In: SENSORS, 2013 IEEE. IEEE, pp 1–4
25. Tasnim R, Khan S, Arshad A, Hussein MR (2018) Non-contact capacitive technique for biomass flow sensing. Indones J Electr Eng Comput Sci 11(2):531–541. Scopus
26. RumanaTasnim, Sheroz Khan, MusseMohamud, &Atika Arshad. (2016). A qualitative analysis of biomass flow sensing behavior using capacitive technique. IIUM Eng J 17(1):29–40. Scopus
27. Tasnim R, Khan S, Arshad A (2016) Assessing the solid mass flow sensing behaviour through capacitive sensing technique. In: IEEE student conference on research and development (SCOReD), 13–14 December, 2016. Putrajaya, Malaysia. IEEE, pp 1–4
28. Al-Khateeb KAS, Tasnim R, Khan S, Mohammod M, Arshad A (2013) Experimental evaluation of agricultural biomass flow sensing behaviour using capacitive technique. In: IOP conference series: materials science and engineering, 2–4 July, 2013. Kualalampur, Malaysia, vol 53, no 1. IOP Publishing, p 012034
29. do Nascimento Wrasse A, dos Santos EN, da Silva MJ, Wu H, Tan C (2022) Capacitive sensors for multiphase flow measurement: a review. IEEE Sens J 22(22):21391–21409
30. Tasnim R, Khan S, Mohamed M, Arshad A, Saquib N (Jul 2012) A qualitative overview of capacitive sensing techniques for two phase flow measurement. In: International conference on computer and communication engineering (ICCCE), 3–5 July, 2012. Kualalampur, Malaysia. IEEE, pp 190–194
31. Arshad A, Kadir KA, Khan S, Alam AZ, Tasnim R (2015) A low cost capacitive sensing system for identifying and detecting falls of senior citizens. In: 2015 IEEE 3rd international conference on smart instrumentation, measurement and applications (ICSIMA). IEEE, pp 1–5
32. Arshad A, Khan S, Alam AZ, Tasnim R, Gunawan TS, Ahmad R, Nataraj C (2016) An activity monitoring system for senior citizens living independently using capacitive sensing technique. In: 2016 IEEE international instrumentation and measurement technology conference proceedings, 23–26 May, 2016. Taipei, Taiwan. IEEE, pp 1–6
33. Arshad A, Khan S, Alam AZ, Tasnim R, Boby RI (2016) Healthand wellness monitoring of elderly people using intelligent sensing technique. In: International conference on computer and communication engineering (ICCCE), 26–27 July, 2016. Kuala Lumpur, Malaysia. IEEE, pp 231–235
34. Arshad A, Alam AHMZ, Khan S, Tasnim R (2015).Passive monitoring techniques for elderly: A review on recent development and prospective. In: Proceedings of the World Congress on Engineering and Computer Science, vol 1, 21–23 October, 2015, Sam Francisco, USA. Scopus
35. Rajala S, Siponkoski T, Sarlin E, Mettanen M, Vuoriluoto M, Pammo A, Tuukkanen S et al (2016) Cellulose nanofibril film as a piezoelectric sensor material. ACS Appl Mater & Interfaces 8(24):15607–15614

36. Akdogan EK, Allahverdi M, Safari A (2005) Piezoelectric composites for sensor and actuator applications. IEEE Trans Ultrason Ferroelectr Freq Control 52(5):746–775
37. Shen Z, Tan CY, Yao K, Zhang L, Chen YF (2016) A miniaturized wireless accelerometer with micromachined piezoelectric sensing element. Sens Actuators A: Phys 241:113–119
38. Arpin-Pont J, Gagnon M, Tahan AS, Coutu A, Thibault D (2015) Methodology for estimating strain gauge measurement biases and uncertainties on isotropic materials. J Strain Anal Eng Des 50(1):40–50
39. Ahmed H, Liu Y, Selami Y, Tao W, Zhao H (2019) Development-of-a-real-time-axial-force-measurement-system-and-IoT-based-monitoring-for-smart-bearing. https://doi.org/10.5281/zenodo.3566337
40. Ekreem N, Olabi AG, Prescott T, Rafferty A, Hashmi, M.S.J. (2007) An overview of magnetostriction, its use and methods to measure these properties. J Mater Process Technol 191:96–101. https://doi.org/10.1016/j.jmatprotec.2007.03.064
41. Aghili F, Buehler M, Hollerbach J (2001) Design of a hollow hexaform torque sensor for robot joints. Int J Robot Res 20:967–976
42. Yeh CT, Tsai NC, Chiu HL, Sue CY (2014) Digital high-resolution torque sensor and signal processing. IEEE Sens J 15(4):2093–2100
43. Mohankumar P, Ajayan J, Yasodharan R, Devendran P, Sambasivam R (2019) A review of micromachined sensors for automotive applications. Measurement 140:305–322
44. Marciszko F (2004) Torque sensor-based powertrain control
45. Adan S, Sharma S (Aug 2018) Innovative technologies for electrical domain. Int J Trend Sci Res Dev (IJTSRD) 2(5):2384–2387. ISSN: 2456-6470
46. Reznicek M, Szendiuch I, Reznicek Z (2010) Thermodynamic sensors new opportunities for measuring and control in industrial applications. In: 3rd electronics system integration technology conference ESTC, Berlin, Germany, pp 1–5. https://doi.org/10.1109/ESTC.2010.5642802
47. Gidik H, Bedek G, Dupont D, Codau C (2015) Impact of the textile substrate on the heat transfer of a textile heat flux sensor. Sens Actuators A: Phys 230:25–32
48. Khoshnoud F, de Silva CW (2012) Recent advances in MEMS sensor technology—thermo-fluid and electro-magnetic devices. IEEE Instrum Meas Mag 15(3):16–20
49. Hu W, Morris K, Zhang Y (Dec 2016) Sensor location in a controlled thermal fluid. In: 2016 IEEE 55th conference on decision and control (CDC). IEEE, pp 2259–2264
50. Ali MM, Islam R, Lim KS, Gunawardena DS, Yang HZ, Ahmad H (2015) PCF-cavity FBG Fabry-Perot resonator for simultaneous measurement of pressure and temperature. IEEE Sens J 15(12):6921–6925
51. Alt A (2021) Flow sensor solutions in modern medical ventilators. [Online] Available: Wu J (2018) A basic guide to thermocouple measurements. Texas Instruments application report
52. Masek P, Zeman P, Kolar P (2021) Cutting temperature measurement in turning of thermoplastic composites using a tool-work thermocouple. Int J Adv Manuf Technol 116:3163–3178
53. Wei C, Li L (2020) Thermometry in laser micro/nanofabrication
54. Thamrin M, Subandi A, Idris I, Ahmad A (2007) Experimental characterization of a thermopile with aluminum-n-type polysilicon junctions as a base for multi-directional flow sensor implemented in a silicon chip. ITB J Eng Sci 39:98–108. https://doi.org/10.5614/itbj.eng.sci.2007.39.2.2
55. Burton D, Sheridan J (2011) thermistor-based instrument for measuring vehicle cooling airflow. In: Vehicle thermal management systems conference and exhibition (VTMS10). Woodhead Publishing, pp 145–154. ISBN 9780857091727
56. Zhang JXJ, Hoshino K (2019) Chapter 6—Mechanical transducers: cantilevers, acoustic wave sensors, and thermal sensors. In: Zhang JXJ, Hoshino K (eds) Micro and nano technologies, molecular sensors and nanodevices, 2nd ed. Academic Press, pp 311–412. ISBN 9780128148624
57. Christenson J (2019) Chapter three - sensors and transducers. In: Segil J (ed) Handbook of biomechatronics, Academic Press, pp 61–93. ISBN 9780128125397

58. Das S, Sarkar TS, Chakraborty B (2018) Simple approach to design a capacitive rotary encoder. IET Sci Meas Technol 12(4):500–506
59. Sandin P (2003) Robot mechanisms and mechanical devices illustrated. McGraw Hill Professional
60. Soriano Heras E, Blaya Haro F, De Agustin Del Burgo JM, Islán Marcos M, D'Amato R (2018) Filament advance detection sensor for fused deposition modelling 3D printers. Sensors 18(5):1495
61. Arrigoni M, Monchalin JP, Blouin A, Kruger SE, Lord M (2008) Laser Doppler interferometer based on a solid Fabry-Perot etalon for measurement of surface velocity in shock experiments. Meas Sci Technol 20(1):015302
62. Moir CI (May 2009) Miniature laser Doppler velocimetry systems. In: Optical sensors 2009, vol 7356. SPIE, pp 122–133
63. Gomes DF, Luo S (2022) GelTip tactile sensor for dexterous manipulation in clutter. In: Tactile sensing, skill learning, and robotic dexterous manipulation. Academic Press pp 3–21
64. Miguel J (2020) Sensing using light: a key area of sensors. Sensors 21(19):6562
65. Liu J, Liang Y, Wang L, Wang B, Zhang T, Yi F (2016) Fabrication and photosensitivity of CdS photoresistor on silica nanopillars substrate. Mater Sci Semicond Process 56:217–221
66. Adams-McNichol AL, Shiell RC, Ellis DA (2019) Accurate, photoresistor-based, student-built photometer and its application to the forensic analysis of dyes. J Chem Educ 96(6):1143–1151
67. Argeseanu A, Ritchie E, Leban KRISZTINA (2009) A new solar position sensor using low cost photosensors matrix for tracking systems. WSEAS Trans Power Syst 6(4):189–198
68. Sun D (Nov 2020) Optimized design of photodiode amplifier circuit and its application in gas sensing. In: AOPC 2020: optical sensing and imaging technology, vol 11567. SPIE, pp 374–378
69. Driesse A, Zaaiman W, Riley D, Taylor N, Stein JS (Jun 2016) Investigation of pyranometer and photodiode calibrations under different conditions. In: 2016 IEEE 43rd photovoltaic specialists conference (PVSC). IEEE, pp 0127–0132
70. UllaH H, KHEra E, MaliK M (2019) Obtaining the optical properties of chicken stomach and blood with laser-based diffuse-wave spectroscopy (DWS): An Ex vivo study. Lasers in Engineering (Old City Publishing), p 43
71. Zimmermann H (2023) Ultra-sensitive PIN and avalanche photodiode receivers. IOP Publishing
72. Shahriar KMF (2021) Automatic headlight controlling of vehicle using ambient light sensor based on phototransistor. ScienceOpen preprints
73. Wang F, Wang Z, Jiang C, Yin L, Cheng R, Zhan X, He J (2017) Progress on electronic and optoelectronic devices of 2D layered semiconducting materials. Small 13(35):1604298
74. Elzwawy A, Pişkin H, Akdoğan N, Volmer M, Reiss G, Marnitz L, Schmalhorst JM (2021) Current trends in planar Hall effect sensors: evolution, optimization, and applications. J Phys D: Appl Phys 54(35):353002
75. Stiawan R, Kusumadjati A, Aminah NS, Djamal M, Viridi S (Apr 2019) An ultrasonic sensor system for vehicle detection application. In: Journal of physics: conference series, vol 1204, no 1. IOP Publishing, p 012017
76. Karimi M, Babaians E, Oelsch M, Steinbach E (2021) Deep fusion of a skewed redundant magnetic and inertial sensor for heading state estimation in a saturated indoor environment. Int J Semant Comput 15(03):313–335
77. Zadeh NRN, Cruz JCD (Nov 2016) Smart urban parking detection system. In: 2016 6th IEEE international conference on control system, computing and engineering (ICCSCE). IEEE, pp 370–373
78. Datta SK (2013) Ultrasonic and electromagnetic NDE for structure and material characterization engineering and biomedical applications
79. Aliew F (2022) An approach for precise distance measuring using ultrasonic sensors. Eng Proc 24(1):8
80. Mishra A, Cleveland RO (2023) Pulsed waves for medical therapy. Power Ultrason, 599–616

81. Hossain N, Al Mahmud MZ, Hossain A, Rahman MK, Islam MS, Tasnim R, Mobarak MH (2024) Advances of materials science in MEMS applications: a review. Results Eng, 102115
82. Bogue R (2013) Recent developments in MEMS sensors: a review of applications, markets and technologies. Sens Rev 33(4):300–304
83. Mohniuc R, Zaicenco A, Kuendig C, Kurmann T (2008) Mems-based data logger for seismic arrays and structural health monitoring. In: Harmonization of seismic hazard in vrancea zone: with special emphasis on seismic risk reduction. Springer Netherlands, pp 309–317
84. Gross JN, Gu Y, Rhudy MB, Gururajan S, Napolitano MR (2012) Flight-test evaluation of sensor fusion algorithms for attitude estimation. IEEE Trans Aerosp Electron Syst 48(3):2128–2139
85. Mirabadi A, Mort N, Schmid F (Dec 1996) Application of sensor fusion to railway systems. In: 1996 IEEE/SICE/RSJ international conference on multi-sensor fusion and integration for intelligent systems (Cat. No. 96TH8242). IEEE, pp 185–192
86. Elmenreich W (May 2007) A review on system architectures for sensor fusion applications. In: IFIP international workshop on software technolgies for embedded and ubiquitous systems. Springer Berlin Heidelberg, Berlin, Heidelberg, pp 547–559
87. Escudier M, Atkins T (2019) A dictionary of mechanical engineering. Oxford University Press
88. Aranjo B, Soori PK, Talukder P (Jun 2012) Stepper motor drives for robotic applications. In: 2012 IEEE international power engineering and optimization conference Melaka, Malaysia. IEEE, pp 361–366
89. Huang X, Goodman A, Gerada C, Fang Y, Lu Q (2011) Design of a five-phase brushless DC motor for a safety critical aerospace application. IEEE Trans Industr Electron 59(9):3532–3541
90. Ripert M (2005) Calibration of analog sensors for the alignment of muon chambers in the CMS experiment (Doctoral dissertation, Florida Inst. Tech.)
91. Ali WH, Kolawole ES, Obiomon P, Fuller JH, Ali S, Cofie P (2016) Rapid prototype with field gate (a design and implementation of stepper motor using FPGA). Circuits Syst 7(8):1392–1403
92. Barkas DA, Ioannidis GC, Psomopoulos CS, Kaminaris SD, Vokas GA (2020) Brushed DC motor drives for industrial and automobile applications with emphasis on control techniques: a comprehensive review. Electronics 9(6):887
93. Boldea I (2018) Electric generators handbook-two volume set. CRC Press
94. Hilal M, Alrikabi HTS, Aljazaery I (2023) A control system of DC motor speed: systematic review. Wasit J Comput Math Sci 2(1):93–111
95. Chung SK, Kim HS, Kim CG, Youn MJ (1998) A new instantaneous torque control of PM synchronous motor for high-performance direct-drive applications. IEEE Trans Power Electron 13(3):388–400
96. Pranata I, Nuryadi S (2019) Rancang Bangun Mesin Cetak Pcb Berbasis Arduino Menggunakan Metode Cnc (Computer Numerical Control) (Doctoral dissertation, University of Technology Yogyakarta)
97. Breidi F, Garrity J, Lumkes J (2017) Design and testing of novel hydraulic pump/motors to improve the efficiency of agricultural equipment. Trans ASABE 60(6):1809–1817
98. Xu B, Shen J, Liu S, Su Q, Zhang J (2020) Research and development of electro-hydraulic control valves oriented to industry 4.0: a review. Chin J Mech Eng 33:1–20
99. Swapna P Importance of hydraulics and its role in mechanical and automobile engineering
100. Hegde GS (2010) Mechatronics. Jones & Bartlett Learning
101. Cao Y, Cheng L, Chen XB, Peng JY (2012) An inversion-based model predictive control with an integral-of-error state variable for piezoelectric actuators. IEEE/ASME Trans Mechatron 18(3):895–904
102. Kaltenbacher M (2007) Numerical simulation of mechatronic sensors and actuators, vol 2. Springer, Berlin
103. Bishop RH (2017) Mechatronics: an introduction. CRC Press

104. Cuéllar JHDA (2018) Real-time monitoring system for a building-like structure using wireless sensors. Thesis (Doctoral dissertation, Fachhochschule Aachen)
105. Vyatkin V, Hanisch HM, Pang C, Yang CH (2008) Closed-loop modeling in future automation system engineering and validation. IEEE Trans Syst Man Cybern Part C (Appl Rev) 39(1):17–28
106. Dupont B (2020) Touchscreen-friendly fingertip system for tactile feedback in prosthetic hands (Doctoral dissertation, University of British Columbia)
107. Krzysztofik I, Blasiak S (2024) Analytical solutions of the 3-DOF gyroscope model. Electronics 13:1843. https://doi.org/10.3390/electronics13101843

Control System in Mechatronics Engineering

Farhan Mahbub, Enamul Hoq, Rezwan Us Saleheen, Ahmed Farhan, Md. Tarek Ur Rahman Erin, and Mustafizur Rahman

Abstract Control systems in mechatronics play a significant role in regulating and managing the performance of mechatronic systems, which are integrated systems that combine mechanical, electronic, computer, and control engineering components. The required performance, stability, and behavior of mechatronic systems must be maintained via the control system. Control Systems with advanced applications provide fundamental and cutting-edge engineering knowledge on how control and robotics technologies are applied in the life sciences. Based on the recent developments, the promising field of control systems and robotics in bioengineering is likely to lead to ground-breaking innovations in the following decade. Although control engineering and mechatronics engineering undergraduate and graduate students are the target audience for this chapter, academics and practitioners looking to improve their quantitative comprehension of physiological processes may also find it useful.

Keywords Control · Performance · Stability · Robotic technologies · Innovations · Mechatronics

1 Introduction

Human interactions with machines are shifting as a result of the rapid advancements in mechatronics and control systems. Mechatronics is a field that combines computer science, electronics, and mechanical engineering to develop and manage complex systems. Control systems, which keep an eye on and manage the actions of machinery, are getting more and more complex. New developments in mechatronics and control systems enable the development of more economical, dependable, and effective solutions for a range of uses. For instance, mechatronics and control systems are used in

F. Mahbub (✉) · E. Hoq · R. U. Saleheen · A. Farhan · M. Rahman
Department of Mechatronics Engineering, World University of Bangladesh, Dhaka, Bangladesh
e-mail: mahbub3@mte.wub.edu.bd

Md. T. U. R. Erin
Department of Mechanical Engineering, Textile Engineering College, Chattagram, Bangladesh

M. M. Rahman et al. (eds.), *Mechatronics*, Emerging Trends in Mechatronics,
https://doi.org/10.1007/978-981-97-7117-2_3

the automobile sector to produce self-driving autonomous vehicles. Robotic surgery systems and sophisticated prostheses are developed in the medical industry through the application of mechatronics and control systems. Figure 1 shows the implementation of control system for mechatronics application. The manufacturing process is impacted by mechatronics and control systems, which result in increased efficiency and cost savings in addition to improved safety and dependability. Mechatronics and control systems are designed to increase productivity by automating processes. Businesses that automate processes can reduce labor costs and increase production rates. This could lead to increased profits and a competitive edge in the industry. Businesses can save labor costs by automating processes. Moreover, mechatronics and control systems can help reduce energy costs by optimizing energy use. Mechatronics and control systems can reduce the likelihood of human mistake, which can improve worker safety. Additionally, automated systems can lessen the possibility of mishaps and injuries. Because automated mechatronic systems are built to be dependable and consistent, they can minimize downtime and guarantee that operations are carried out without a hitch. But there are also certain difficulties. Control systems and mechatronics face a wide range of difficulties. As technology advances, so do the challenges involved in designing, building, and maintaining mechatronics and control systems. One of the main challenges is the complexity of the systems. In order to accomplish the intended result, a combination of mechanical, electrical, and computer components is used in mechatronics and control systems. The danger of failure, cost, and design time may all rise as a result of this complexity.

The requirement for precise and trustworthy sensing and control presents another difficulty. In safety–critical applications, where a failure could have dire repercussions, this is extremely crucial. Sophisticated algorithms are necessary for accurate and dependable sensing and control, but they can be challenging to create and maintain. Finally, given how quickly technology is advancing, it could be difficult to keep up with the most recent developments in mechatronics and control systems. It can be challenging to keep up with the most recent developments in technology, as new

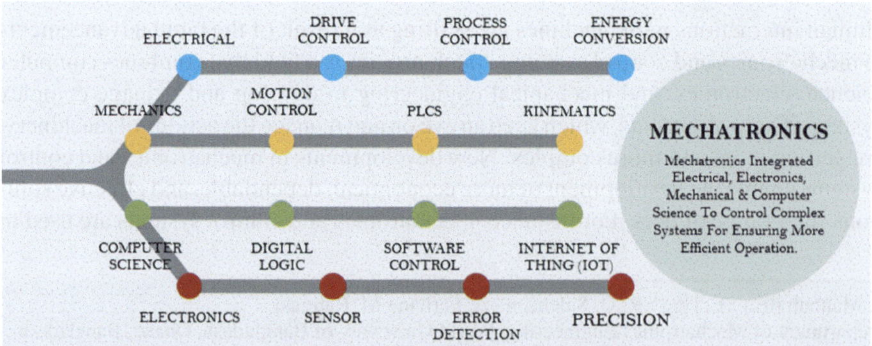

Fig. 1 Implementation of control system in the field of mechatronics

ones are always being developed. This may result in systems that are out-of-date, ineffective, or unable to benefit from recent technology advancements.

Control systems are essential to the design of mechatronic applications such as chain mechanisms, engine control, active suspension, plant modeling, and automatic parking and temperature monitoring systems. Robot design was made possible by the use of an Arduino microcontroller and embedded system for programming. System design benefits from having the sensor, controller, and actuator—the main parts of a category control system. As the so-called "brain" of the mechatronic system, the controller processes inputs from sensors and users to produce commands that are sent to the actuator. Devices called actuators have the ability to transform electrical energy into mechanical energy. For this purpose, the technology that comes from the term "control systems" is the most important. Humans now create devices and systems that work for us and have the fewest errors imaginable. A control system is a tool used to maintain or change any amount of interest in any domain in the way that is wanted. The terms "plant," "process," and "controlled system" are used to refer to the various systems that need to be managed. All of these theories combined are referred to as unified control theory. Servomechanism was brought to process control earlier, and cybernetics theory followed.

1.1 Types of Control Systems

Control systems fall into two primary categories: closed loop and open loop. An open-loop control system is one whose output remains constant regardless of the actual conditions in which it functions. This suggests that the input and the output are not compared, or that feedback is not included in this system. An illustration of an open-loop control system is shown in Fig. 2.

A time-based design in an open-loop control system is represented by the bread toaster. It might be challenging to toast bread to the ideal golden brown. That being stated, measuring the toast's length is the only way to determine the appropriate benchmark. Toast is supplied according to the demands of the customer. Because of this, the consumer is dependent on trial and error to ensure that the bread gets toasted to their liking. Since open-loop control systems are unable to adapt for interruptions,

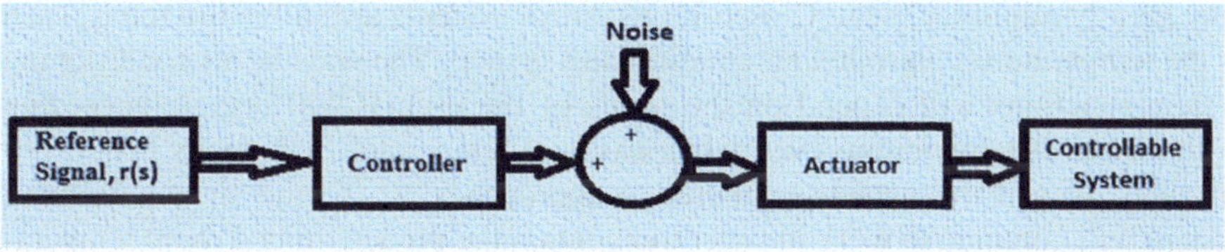

Fig. 2 Open-loop control system

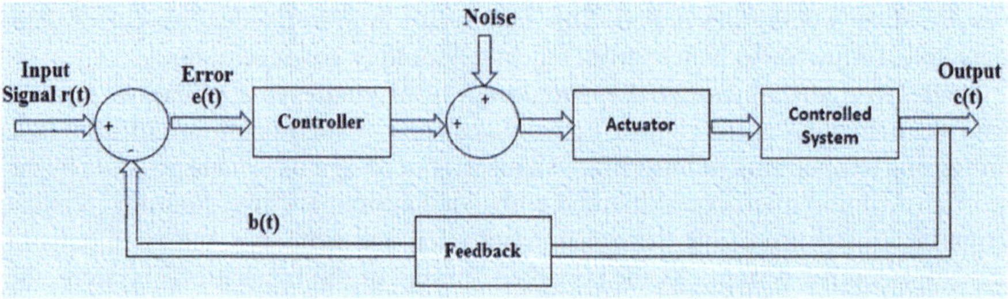

Fig. 3 Closed-loop control system

they are typically not appropriate. As a result, the product won't satisfy the standards. As a result, complex systems are not a good fit for the open-loop control approach.

Feedback control systems are another name for closed-loop control systems. This type of control system aims to address system flaws by utilizing one or more feedback loops or circuits. This implies that in order to make sure the derived output is established even before it is modified by the controller, the system's output will be watched over and compared to the input. Because it acts as the set-point at the input for system testing, the input signal, r(t), is also referred to as the reference signal. Because the b(t) signal is sent to the comparator's negative sign in a closed-loop control system like the one in Fig. 3, which is a negative feedback control mechanism.

For such systems, the error can be expressed using Eq. (1):

$$e(t) = r(t) - b(t) \tag{1}$$

A system that supplies the b(t) signal to a positive comparator follows a positive feedback mechanism. Equation (2) given below can be utilized to determine the error for such a system.

$$e(t) = r(t) + b(t) \tag{2}$$

The resultant error signal, e(t), serves as the input controller. In order to guarantee that the issue can be fixed, the controller will take action. Depending on the type of controller being utilized, faults can be eliminated more successfully. Certain controllers have the ability to completely remove faults, while others can only greatly minimize them. Closed-loop control systems are still in use today, such as the temperature-controlled air conditioning system. The way the air conditioning system functions will depend on the needs of the user. It will operate according to the user-set temperature even when external temperature variations occur. For instance, even if the outside temperature approaches 30 °C, the air conditioning system will continue to work in order to maintain the predetermined room temperature of 18 °C. Other examples of closed-loop control systems are processing control systems, missile launchers, voltage stabilizers, and other systems.

1.2 Control System Components and Element

A mechanical system with some flaw in its design will never work properly, even with the addition of a sophisticated control system. In order to provide a unified model of a mechanical system, its dynamics, and its control system structure, mechatronic systems must be modeled, simulated, and integrated (it is ideal that (sub-) models be reused). This will make it easier to analyze, predict system effects, and perform better overall. It will also generally result in a more flexible system or lower system costs [1]. The control unit (physical controller) and the selection, design, and synergistic integration of the control algorithm subsystems are two of the most crucial decisions made during the mechatronics design process. The system, algorithm, and electronic unit subsystems are identified as the three main subsystem components of this level. Six fundamental types of programmable control systems can be used to organize the physical-controller subsystem: Microcomputers, digital signal processors (DSP), programmable logic controllers (PLCs), microcontrollers (Mc), and application-specific integrated circuits (ASICs) [2]. Adaptive, intelligent (fuzzy logic, neural networks, expert systems, and genetic), feedforward with adaptive features, ON–OFF, PID modes, and feedforward with adaptive features are some more control techniques. A number of significant variables may influence the control unit or algorithm selection: Programming language (e.g., C + +) and programming environment (e.g., how control executes program) impact machine development time and operation. (a) Operation (processing tasks and instructions); (b) Strength/environment (industrial, soft, etc.); (c) Longevity (easy replace/repair over controller life and availability of replacement components); (d) Security (viruses, etc.); (e) the application's safety criticality and the amount of items to be produced; (f) as well as the unit cost, end product cost, accuracy, and time to market required; (g) size-space saving; (h) Combination; and (i) Power [3, 4].

The initial stage of designing a mechatronics system is choosing the control unit, algorithm, sensors, and actuators. The electronic unit subsystem, which includes interface and signal conditioning, is then selected and integrated. These subsystems include power supply, drive circuits, data collection systems, and signal processing-conditioning circuits. They are chosen with the goals of matching component ratings and standards, maximizing system performance, and facilitating communication with every system component. The transfer of sensor signals to the control unit and signals from the control unit to the actuator are made possible by interface hardware, which also enables analog/digital interface. This can be ascribed to the reality that control units are digital yet the majority of measuring and control equipment is analog. To support the actuator subsystem, the electronic unit must modify both the actuator unit and the control signals it generates to the appropriate level. At times, this gadget can perform extra functions including system protection, functional parameter setting, and display. Basic electronic parts including transistors, potentiometers, condensers, rectifier diodes, stabilizers, and resistors are found in the electronic unit.

b) Linear integrated circuits, such as current generators, ramp generators, demodulators, adders, PID regulators, and oscillators that are built on operational amplifiers.
c) A/D and D/A convertors as supply units [3, 4].

1.3 Importance of Control System in Mechatronics Engineering

Advances in information technology and decision-making, together with the collaborative integration of various key engineering domains, are contributing to the growing complexity, diversity, and multifarious nature of engineering issues. Multidisciplinary engineering systems must be used to solve multidisciplinary challenges; these modern interdisciplinary systems are referred to as mechatronic systems. A mechatronics engineer must possess strong modeling, simulation, analysis, and control skills in addition to a thorough understanding of the fundamental issues with hardware implementation [1, 2]. The mechatronics design process can become exceedingly complicated due to the involvement of numerous disciplines, presenting daunting hurdles for engineering instructors. It is expected of mechatronics engineers to develop products with a focus on functionality, reduced costs, speed, accuracy, efficiency, and synergy in order to meet various restrictions.

Mechatronics design integrated concepts composed of a number of methodical, clear-cut, and simple processes. The selection, design, and integration of the control unit and algorithm subsystems are two interconnected subsystems in the mechanical system design process. Mechatronics systems are intended to be designed with synergy and integration toward constraints like higher performance, speed, precision, efficiency, lower costs, and functionality. This allows them to function with exceptionally high levels of accuracy and speed despite the negative effects of system nonlinearities, uncertainties, and disturbances.

1.4 Benefits of Control System

- **Fast and Error-Free**: Control systems operate more quickly than humans and respond to inputs very quickly. They accomplish the same task in less time and with less human labor. Control systems produce extremely little error, which makes them ideal for tasks where errors and mistakes are highly likely to occur. Because they are made to be precise and function without error, they subsequently lessen the mistakes that humans are prone to making.
- **Accuracy**: The output of the control system is entirely reliable. With the same input, they always yield the same result because of their extraordinary accuracy. Productivity is increased and output variance is reduced by the control system.

- **Adaptable and Efficient**: It is possible to program the control system to detect changes and react appropriately to them. They practically never make mistakes and can operate in any kind of setting. They work faster, more accurately, and with greater precision, making them efficient.
- **Security**: The control system reacts more precisely while producing the same result. With such amazing precision, the control system decreases output variance while increasing productivity. Control systems provide a variety of monitoring features that allow them to keep tabs on the behavior of the system. If they notice any irregularities or unusual activity, they alert the user or stop the process. The prompt answer may result in the temporary termination of that particular process. For safety reasons, automatic anomaly detection in the control system is essential.
- **Optimal Resource Utilization**: Control systems are able to maximize and make the best use of the resources that are made available to them. It improves resource utilization and reduces the need for human intervention, especially in challenging jobs. For instance, if production changes in the agriculture sector, the control system might modify supply and demand to avoid overproduction of the goods.
- **Real-Time Monitoring System**: Real-time observation from the control system facilitates prompt decision-making when assessing system performance. Additionally, it enables the detection of abnormalities in the system, and early mistake detection can minimize a significant potential problem. For the manufacturing industry, the parameters can be adjusted in reaction to modify the surrounding environments.
- **Integration with Other Complex Systems**: It improves performance by integrating its function with other systems. The system's capabilities are increased through integration with other systems, and cooperation within the system ensures efficiency and production.
- **Optimization**: It guarantees the optimal path of action to reach the intended result by breaking down the complex process into smaller, distinct segments. The capacity of the control system to monitor and control remotely is one of its most crucial aspects. This feature guarantees error-free performance, especially in tasks where errors are prone to happen.

2 Mathematical Preliminaries

2.1 Laplace Transform

One of the most important mathematical tools with widespread use in many engineering disciplines is the Laplace transform. This kind of transform was initially employed by the French mathematician "Pierre Simon De Laplace" in his work on "Probability Theory." These transforms are the steps associated with a mathematical process known as the Laplace transform. Such methods are used to "transform" a variable (in space, or at time) to a parameter (s), under some criteria. One variable is changed at a time by it. By substituting new variables for the original ones that represent their Integral and Derivative counterparts, the Integral–Differential Equations can be converted to Algebraic Equations using the Laplace Transform, a helpful tool for studying any electrical circuit [5].

Since one cannot afford to utilize the trial-and-error method on a major engineering project, simulation and design are the critical stages in the physical implementation of any idea in engineering. It is never possible to solve them in the time domain, differential equations in particular. Hence, translate these intricate time domain equations into the frequency domain, where they become easily solvable algebraic functions, to simplify the situation. To put it simply, the Laplace Transform will act as a translator for the visitor from abroad [6]. A given derivative function with real variable t can be transformed into a complex function with variable s using the Laplace transform, which applies an integral transform. Assuming $t \geq 0$, let's suppose that the function f(t) complies with a few requirements, which will be discussed in more detail later. Any function f(t) can have its Laplace transform represented by L{f(t)}, and its Laplace transform formula defines F(s):

$$F(s) = \int_{0}^{+\infty} f(t).e^{-s.t}.dt \tag{3}$$

2.2 *Transfer Functions*

A transfer function illustrates the relationship between the input and output signals in a control system with several different input values. A control system is shown by a block diagram (Fig. 4), in which the various input and output signals are represented by arrows, and blocks stand in for the transfer function. Excitation is the reference input for all control systems. It acts through a transfer operation, sometimes called the transfer function, to produce an effect that results in a regulated output or reaction [6].

In a Laplace transform, the transfer function will be as follows (Eq. 4) where the input is represented by R(s) and the output by C(s).

Fig. 4 A control system

$$G(s) = \frac{C(s)}{R(s)} \Rightarrow R(s).G(s) = C(s) \quad (4)$$

Alternatively, the system output function is obtained by multiplying the transfer function by the input function. If all initial conditions are zero, the transfer function of a control system is simply defined as the ratio of the output variable's Laplace transform to the input variable's Laplace transform.

2.3 *State-Space Representation of Dynamic Systems*

The interaction among the system's input and output is demonstrated by the transfer function. The equation, which simplifies the process of determining the system's behavior, is the main advantage of this revelation. The conditions of a system and how they change over time constitute its state space. In this case, the system's output variables depend on how the input and state values work. The state matrix, the vector that represents the output of the model, reveals the interaction between the variables determining the behavior of the system [7]. This description makes it feasible to see the system's current state, which occasionally makes it possible to predict how it will behave in the following time step. The output vector y(t) refers to the state vector u(t) that produces how the system responds by showing the state vector x(t)'s present condition as a result of the input factors u1, u2,..., um. It is possible to characterize a system of n first-order differential equations as a system having m inputs and r outputs. Equation 5, the state equation, below, represents it:

$$\frac{dx(t)}{dt} = f(t, x(t), u(t) \quad (5)$$

The system's input values $x(0) = x^S$ and state variable functions determine the system's output values when the initial conditions are satisfied. Equation 6, which represents the system's output, is as follows:

$$y(t) = g(t, x(t), u(t)) \quad (6)$$

Here:

$x^T = (x_1, x_2, \ldots, x_n)$; state variables vector,
$u^T = (u_1, u_2, \ldots, u_m)$; input variables vector,
$y^T = (y_1, y_2, \ldots, y_r)$; output variables vector,
$f^T = (f_1, f_2, \ldots, f_n)$; nonlinear vector functions,
n = Number of state variables,
m = Number of inputs,
r = Number of outputs.

Equation (5) is referred to as the dynamics equation, and Eq. (6) is its output equation. The continuous nonlinear time variant system can be demonstrated by these two equations. The system representation is given as follows if the functions f and g are not explicitly dependent on time t. The time invariant systems are represented by Eqs. 7 and 8; they are characterized by a constant across time.

$$\frac{dx(t)}{dt} = F(x(t), u(t)) \tag{7}$$

$$y(t) = g(x(t), u(t)) \tag{8}$$

2.4 *Laplace Transform and Its Application in Control Systems*

The Laplace transform has numerous uses in control systems [8]. This transform is applied, for instance, to control system stability analysis. Additionally, the system response to specified initial conditions and input signals is computed using this transform.

i. Laplace Transform Basics

A mathematical technique called the Laplace transform can make control system analysis and design simpler. It has the ability to translate intricate differential equations that represent a system's dynamic activity into more straightforward algebraic equations that represent a system's frequency response. A system's response to various input signals, like as sinusoids, impulses, or steps, is measured by its frequency response. Additionally, the Laplace transform can be used to create controllers that enhance a system's performance and stability, such as proportional-integral-derivative (PID) controllers. Complex differential equations are reduced to a simpler form based on polynomials by its application.

2.4.1 Laplace Domain Analysis

One technique for evaluating and verifying the control systems design that makes use of the Laplace transform is Laplace domain analysis. As a result, the transfer function of any system needs to be calculated, which is equal to the input divided by the Laplace transform ratio. The transfer function may provide important information about the corresponding system, such as its phase, gain, zeros, and poles. Many methods, such as block diagram reduction, signal flow graphs, and state-space representation, can be used to find the transfer function. For Laplace domain analysis, software programs like Simulink or MATLAB may be utilized.

2.4.2 Time Domain Analysis

Time domain analysis is another technique you may use to verify and validate your control systems design utilizing Laplace transform. Finding the differential equation describing the system's solution, or the system output's inverse Laplace transform, is required for this. With the inverse Laplace transform, one may ascertain the system's responses to different input signals over time, including its rising, settling, overshoot, and steady-state error rates. Many methods such as the residue theorem, convolution theorem, and partial fraction expansion can be applied to determine the inverse Laplace transform. Software tools like MATLAB or Simulink can also be used for time domain analysis.

2.4.3 Frequency Domain Analysis

Frequency domain analysis is the third method for testing and validating control systems design utilizing the Laplace transform. The phase and magnitude expressed in the output response as a result of the applied frequency must be ascertained, which is known as the system's frequency response. From the frequency response, one can deduce the system's bandwidth, resonance, gain margin, and phase margin. Plots such as the Bode, Nyquist, and Nichols can be used to find the frequency response. MATLAB or Simulink are the two kinds of programs that are used in frequency domain analysis.

2.4.4 Experimental Data Collection

Experimental data from any real system or a prototype is needed to support the theoretical study utilizing the Laplace transform. Through comparison with the anticipated outcomes of the theoretical analysis, experimental data can aid in the validation and verification of the control systems design. The system can be configured to receive various input signals in order to collect experimental data. The output signals can then be measured using instruments, sensors, or other data gathering devices. Additionally, experimental data collection and analysis can be done with software tools like Simulink or MATLAB.

2.4.5 Experimental Data Comparison

By comparing the experimental data with the theoretical results, the control systems design employing the Laplace transform and experimental data is finally tested and validated. The data are compared using a variety of techniques, including statistical analysis, correlation analysis, and error analysis. Data comparison is also done using software tools like Simulink and MATLAB. The comparison can assist in determining any potential flaws, restrictions, or enhancements as well as assessing the precision and dependability of the control systems design. Laplace transforms have other uses in the fields of science and engineering.

- It is used to send signals to both ends of the media in the field of communications. For example, signals transmitted over a phone are initially transformed into a time-varying wave, which is then superimposed on the medium.
- Numerous engineering processes, involving digital signal processing, electrical circuit analysis, system design, and controlling load frequencies in power systems, also make usage of it.

3 Modeling and Analysis of Mechatronic Systems

Mechatronics engineers, who design and create complex systems that incorporate mechanical, electrical, and software components, have access to advanced tools such as system modeling and simulation. Mechatronics engineers may increase system performance, decrease errors and hazards, work together more effectively, and save time and money by utilizing these techniques [9]. The procedures of developing and testing virtual representations of actual or hypothetical systems using computer software, mathematical formulas, and physical principles are known as system modeling and simulation. System modeling is the process of using graphical or textual languages, such as state machines, block diagrams, or code, to define the behavior, structure, and relationships of the system's component parts. In order to examine how the system responds to different inputs, situations, and parameters, system simulation entails executing the model on a computer or hardware platform and applying numerical methods. When creating and developing mechatronic systems, system modeling and simulation can be quite helpful. Using this technique, various design options can be explored and enhance system efficiency prior to constructing tangible prototypes. It can also be used to test the system in extreme or realistic situations, as well as to validate and verify system needs and specifications. Furthermore, through system modeling and simulation, clients, partners, and stakeholders can be demonstrated and informed about the behavior and functionality of the system. Finally, by making it simpler to incorporate pre-existing models and code from many domains and sources, it promotes collaboration among disparate teams.

3.1 Modeling Mechanical Systems

Depending on the motion, there are typically two distinct types of mechanical systems [10].

- Translational mechanical systems.

- Rotational mechanical systems.

3.1.1 Modeling of Translational Mechanical Systems

Systems that have been mechanically translated move in a straight line. Three primary components comprise these systems. Mass, spring, and dashpot, or damper, are those three. A translational mechanical system responds to a given force by encountering opposing forces because of its mass, elasticity, and friction. The mathematical sum of the forces acting on the system is zero because the applied force and the opposing forces are acting in opposite ways.

3.1.2 Modeling of Rotational Mechanical Systems

Systems that rotate mechanically revolve around a fixed axis. There are three primary components to these systems. They are dashpot, torsional spring, and moment of inertia. A spinning mechanical system's moment of inertia, elasticity, and friction causes opposing torques to oppose any applied torque. Since the applied torque and the opposing torques work in opposite directions, the mathematical average of the torques acting on the system yields zero.

3.2 Electrical Systems Modeling

Consideration is given to electrical circuits containing inductors, capacitors, and resistors. The rules of Kirchhoff and Ohm control the behavior of such systems [11].

4 Steps for Modeling of Electrical System

- To obtain the differential equations representing electrical systems using resistors, capacitors, and inductors, apply Kirchhoff's voltage law or Kirchhoff's current law.
- Next, simulate the model by forming Transfer Functions from the differential equations that describe it.

4.1 Electro-Mechanical Systems Modeling

Devices including motors, sensors, actuators, and controllers that mix mechanical and electrical components are known as electro-mechanical systems [12]. They are widely employed in many different industries, including manufacturing, robotics, automation, and aerospace. The principles of modeling and simulation for electro-mechanical systems must be utilized in order to design, analyze, and troubleshoot these systems. A mathematical model of the interactions of a system's mechanical and electrical components is created through the process of electro-mechanical system modeling and simulation. Such a model can be used to find possible issues, optimize performance, and test various situations. The simulation is the model's execution with the use of software tools and numerical techniques to produce outputs like tables, graphs, and animations.

4.2 Thermal Systems Modeling

A thermal system is a network that transfers thermal energy among several industrial operators in a production group, or it could serve as a local heating and cooling system for a construction site. Given the high cost of investments associated with thermal systems, it is advantageous to employ various modeling techniques in order to achieve the best possible system design and operational strategies [13].

Among the techniques applied to thermal modeling are the following:

- Optimal control of complex thermal systems.
- Dynamic system simulation.
- Detailed component models.

5 Feedback Control Principle

5.1 Introduction to Feedback Control System

For implementing control systems, feedback is a prominent and practical tool. The feedback cycle is a method by which the system analyzes its output and is capable of adapting its operation in order to accomplish the desired result. Every kind of environmental disruption or modification affects how any control system operates [14]. As a consequence, one signal is returned to the input after being retrieved from the output. The error signal generated after this signal is compared to a reference signal. This error signal is received by the controller, which then modifies the output. A feedback system is the way we term such a system. A positive feedback mechanism represents a system that has a positive feedback signal. The reference input signal in addition to the feedback signal adds up to the error signal in a positive feedback system. On the other hand, the concept of "negative feedback system" refers to a system that generates a negative feedback signal. The error signal in a negative feedback system usually an indication of the discrepancy between the reference input signal and the feedback signal.

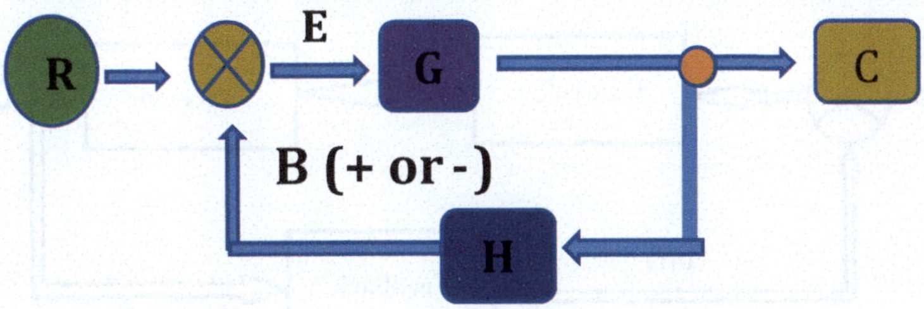

Fig. 5 A Feedback System

The figure below (Fig. 5) has the following labels:

- R = Input response.
- E = Error.
- G = Gain (Forward path).
- H = Feedback Gain.
- C = Output response.
- B = Feedback response.

The following points represent the impacts of feedback on a control system:

- The error among both the input and the output of the system has reduced.
- A reduction in system gain of $1/(1 \pm GH)$ occurs.
- Increased insensitivity, or a decreased reactivity to change.
- A rise in stability occurs.

5.2 *Closed-Loop Control Systems*

Closed-loop control systems involve regulating actions that are contingent on the outputs produced by the system (Fig. 6). Alternatively, the output of the system that applies the input controls these systems. More accurate system output is produced by input that adapts according to the output. Thus, by modifying the closed-loop system's output via a feedback connection, controllability is attained [15].

5.3 *Proportional-Integral-Derivative (PID) Control*

In the field of industrial control, proportional-integral-derivative (PID) control is acquiring significant acceptance as the most frequently used control algorithm in industry. PID controllers are commonly utilized because of their versatility in operating conditions and their user-friendly nature, which simplifies their operation for engineers. As the name suggests, the PID approach makes use of three essential

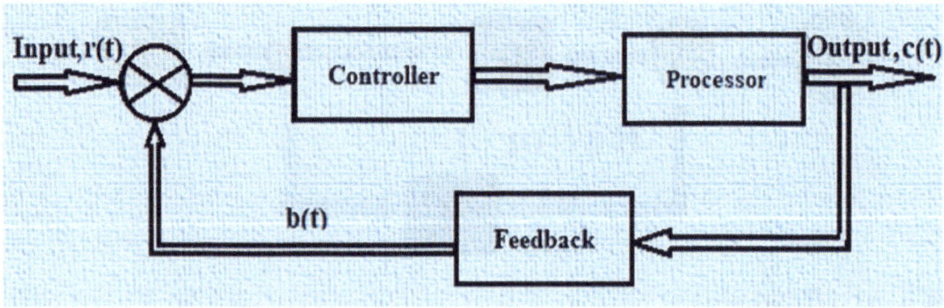

Fig. 6 A closed-loop control system

factors that are changed to yield the optimal result: integral, derivative, and proportional [16]. This kind of control system continuously modifies the output of a process or system to bring it into compliance with a preset point via feedback. Temperature control, flow control, motor control, and many more applications employ PID controllers because they can offer precise and dependable control with a very straightforward implementation.

5.4 Stability Analysis

A system is considered to be stable when its output is under control. It is considered as unstable if not. A stable system yields a bounded output given a bounded input [17]. Based on stability, the systems can be categorized into the categories as follows:

- Absolutely stable system:

A system is considered entirely stable if it remains stable throughout every possible range of values for each of its individual components.

- Conditionally stable system:

A system is regarded as conditionally stable if all of its individual components remain stable within a specific range of values.

- Marginally stable system:

A system can be regarded as marginally stable if, at least for a bounded input, it gives an output signal with a constant magnitude and oscillation frequency.

Routh-Hurwitz, one sufficient and required requirement for stability is known as the stability criteria. A control system can be deemed unstable if it does not satisfy the necessary requirements. Even so, if the control system satisfies the necessary requirement, it might or might not be stable. As a result, the appropriate condition contributes to the assessment of the stability of the control system.

6 Control System Design Technique

6.1 Modern Control Design Approaches

Modern control theory draws on a wide range of study fields as well as a precise set of analysis and synthesis methodologies. Almost every method in control systems theory is based on stability analysis. The use of complex mathematical techniques to govern uncertain, time-varying, and nonlinear systems is the basis of modern control theory. Among these techniques are adaptive control, resilient control, nonlinear control, and optimum control. Modern control theory focuses on state-space domain analysis of system controllability, observability, and optimality using techniques such as Lyapunov functions, Hamilton–Jacobi equations, and Riccati equations. Complex and adaptive controllers that can manage constraints, uncertainties, and system disturbances can be constructed with the aid of modern control theory. Neural network control (NNC), sliding mode control (SMC), and model predictive control (MPC) are a few types of these controllers [18].

Using contemporary control theory, designing a controller involves several steps. It is first necessary to identify the system model by methods like system identification, physical rules, or experimental data. Choosing a suitable controller structure, such as NNC, SMC, or MPC, is the next step. The next step is to determine the controller parameters using analytical or numerical methods such as Lyapunov or Hamilton–Jacobi, or optimization algorithms or linear matrix inequalities. Lastly, the controller performance needs to be verified and tested via simulation or experiments.

6.2 Robust Control Design Techniques

Industries include manufacturing, mining, hardware applications, and the automotive sector all use feedback control systems. In the face of challenging and shifting operating conditions, these control systems are under increasing pressure to function more precisely and better overall due to the growing demands for enhanced efficiency and dependability. New tools and enhanced control theory will be needed by control engineers to design control systems that can fulfill the need for higher performance and resilience when controlling complex processes. Increasing a control system's number of actuators and sensors is a common method of enhancing its functionality. This always results in a control system that is multi-input multi-output, or MIMO. As such, the design process for any modern feedback control system needs to be capable of managing the situation when there are several actuators and sensors [19]. An effective control system performs as planned even when there is a great deal of plant (process) uncertainty. The goal of resilient design is to maintain system performance assurance in the face of faults and model modifications. Systems that are resilient allow their performance to be impacted by outdated or inaccurate models.

A control system is considered robust (represented in Fig. 7) if—

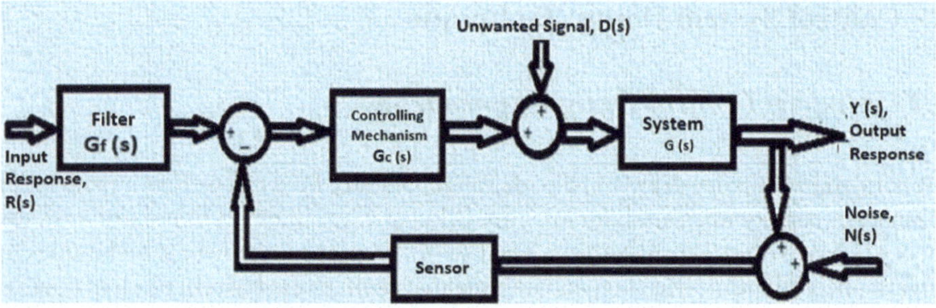

Fig. 7 A robust control system

- its performance meets the specifications even after a set of modifications to the system parameters,
- it has low sensitivities, and
- it is stable throughout a range of parameter fluctuations.

6.3 Optimal Control Strategy

A prerequisite for dynamic systems that meet design goals is optimal control. Control laws that operate in accordance with specified optimality criteria provide optimal control [20]. Several popular approaches for optimum control include the following:

- Linear Quadratic Regulator (LQR)/Linear Quadratic Gaussian (LQG) control.
- Model Predictive Control.
- Reinforcement Learning.
- Extremum Seeking Control.
- H-Infinity synthesis.

Finding an open-loop or closed-loop control function that optimizes a performance metric related to the control and the expected state is the aim of optimum control. For example, an automotive driver would want to reach their destination while also achieving several other goals, such as avoiding obstructions, not driving erratically, and maintaining a safe acceleration level for human passengers. To develop a driving style, these goals might be balanced together. The control designer can provide the desired outcomes and the dynamic model with optimal control, and the algorithm will determine the best course of action. This eases the burden on the designer by letting them focus on what the robot has to achieve rather than how it should accomplish it.

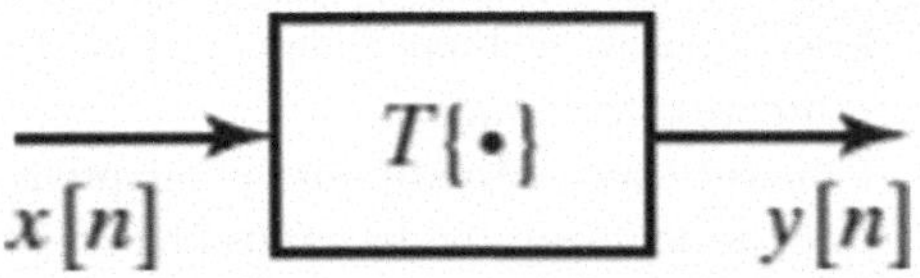

Fig. 8 Graphical representation of discrete-time systems

7 Digital Control Systems

7.1 Discrete-Time Systems

A discrete-time system can be expressed as an operator that converts a series of input values x[n] into a series of output values y[n]. As observed, this is

$$y[n] = T\{x[n]\} \tag{9}$$

Such systems can be shown graphically in Figure (Fig. 8). Procedures for the calculation of the output response values from the input response are represented by an equation.

The input samples x[n] can determine the sequence of output values at every value n for each input value, implying that y at time n could be dependent on any part of the whole sequence x.

8 Advanced Control Techniques in Mechatronics

8.1 Adaptive Control

A subfield of control theory known as "Adaptive Control" focuses on the design and evaluation of control systems that have the ability to alter their behavior in response to external or systemic changes. Adaptive control is an exciting field that is fast developing and has applications in many different fields, including industrial control, robotics, and aerospace. Adaptive control systems have the ability to "learn" from their experiences and modify their control strategies, in contrast to classic control systems, which depend on pre-established models and fixed parameters. They are especially well-suited for handling nonlinear, unpredictable, or time-varying systems due to their adaptability [21].

Key Principles of Adaptive Control:

- **Parameter Estimation:** Accurately determining the system's unknown or changeable characteristics is a major challenge in adaptive control. Online estimating approaches are commonly utilized for this purpose, updating parameter estimations in real time based on input–output data from the system. Recursive least squares, Kalman filters, and gradient-based approaches are a few popular estimation techniques.

- **Control Law Adjustment:** In adaptive control systems, the updated parameter estimations form the basis for changes to the control laws. This adaptation can be made via a variety of methods, such as changing the coefficients in a state feedback controller or the gains in a PID controller. Maintaining the system's current functionality in the face of external changes is the aim.
- **Stability and Convergence:** A vital concern is making sure an adaptive control system is stable. It is necessary to build adaptive controllers in such a way that with respect to every variation of the controller parameters, the system is stable. To ensure optimal control performance, the parameter estimating procedure also needs to converge to the actual system values.

Types of Adaptive Control:

Model Reference Adaptive Control (MRAC) and Self-Tuning Regulators (STR) are the basic kinds of adaptive control.

Model Reference Adaptive Control (MRAC): Using a reference model, the intended system behavior is stated in MRAC. The optimum goal of such a controller is to simulate the system as closely as possible to the reference model. Direct adaptive control, which modifies controller settings in response to variations between the real system and the reference model, is a common technique used by MRAC systems.

Self-Tuning Regulators (STR): Subsequently estimates of the system parameters are used to build the controller employing an indirect adaptive control technique used by STRs. STRs can be particularly beneficial when interacting with systems which require various controller settings for different operating conditions or that have a large number of operating points.

8.2 Fuzzy Logic Control

A unique method for modeling and controlling complex systems is offered by fuzzy logic control, a nonlinear control technique. The initiator of fuzzy logic was Zadeh. This logic is analogous to human thought processes and has many values [22]. It is an extension of traditional set theory with the goal of managing imprecision and uncertainty in data. Fuzzy logic controls functions with approximate or uncertain information, in contrast to standard control techniques that depend on exact mathematical models and crisp values. This makes it especially well-suited for systems with ill-defined or unknown properties. The goal of fuzzy logic is to simplify logic controller design while maintaining control over systems with constant variations. In a fuzzy logic system, the measurement may effectively be partially true; therefore, a fuzzy measurement can range between 0 and 1, when yes equals 1 and no equals 0. The system's rules can be expressed in natural language and then converted to fuzzy logic. For instance, a furnace's design may state "Reduce the amount of fuel used to the furnace if the operating temperature is excessively high." Add more fuel if the furnace continues to operate at a low temperature. Real-world measurements, like the temperature of a furnace, are converted into values between 0 and 1 by identifying their location on a triangle. Typically, the greatest value, or "1," is located at the point of the triangle. Fuzzy sets, which allow for the representation of partial truth or varying degrees of membership inside a set, are fundamental to fuzzy logic control. This is not the case with conventional set theory, which relies on clear-cut set boundaries and binary logic. Fuzzy logic control approximates the behavior of a complicated system by using fuzzy rules, fuzzy inference, and linguistic variables. Making decisions with this method is more instinctive and human-like [23]. Fuzzy logic control has drawn considerable attention in the realm of power electronic systems owing to its capability to handle uncertainties, nonlinearities, and varying operational conditions. Its effective use is evident in a number of power electronic applications, including battery management systems, inverters, motor drives, and DC-DC converters. Compared to traditional control methods, fuzzy logic control offers improved performance, flexibility, and robustness. The following six key assumptions need to be developed while designing a fuzzy control system:

- The availability of the input, output, and state variables must be assumed.
- Presuming the presence of knowledge is necessary.
- The output data set, which can be used to derive rules.
- Assuming that a solution exists is essential.
- Instead of focusing on perfect solutions, engineers should look for "Good Enough" ones.
- A fuzzy logic controller's design needs to be within a suitable range.
- Rather than being explicitly addressed, these issues should be left unaddressed while developing a fuzzy logic controller.

The basic elements of FLCs are as follows:

- The fuzzy value generator, which transforms the input values from crisp to fuzzy.
- About fuzzy input–output relationships, it preserves all of the details. It also has a membership mechanism that specifies which parameters are outcomes for the controlled unit and what variables are inputs for the fuzzy rule base.
- It saves details regarding the course of action of the domain process.
- The FLC's kernel serves as the Inference Engine. Basically, it emulates human decision-making employing approximate reasoning.
- Fuzzy inference engine's fuzzy values are transformed into crisp values by the Defuzzifier.

Advantages of Fuzzy Logic Control

- Designing an FLC is more economical than developing a model-based or any controller when it concerns performance.
- Because FLCs can operate in a wide range of operating circumstances, they are more robust than PID controllers.
- FLCs are adaptable.

- In essence, FLC is made to mimic how humans draw conclusions from their knowledge by applying deductive reasoning.
- Compared to traditional control systems, FLC has higher reliability.
- Using fuzzy logic in a control system results in increased efficiency.

Disadvantages of Fuzzy Logic Control

- FLC must apply a lot of data.
- FLC is not helpful for programs with significantly greater or smaller historical data.
- FLC requires a high level of human competence.
- Requires frequent rule updates.

8.3 Neural Network Control

Promising outcomes have been seen in neural network applications for dynamic system characterization and controls. The multilayered perceptron is a common choice for modeling nonlinear systems and creating general-purpose nonlinear controllers due to its global approximation abilities [24]. There are typically two processes involved in implementing neural networks to control:

- System identification.
- Control design.

A neural network model is created (shown in Fig. 9) during the system identification phase that has to be controlled. The neural network plant concept is utilized to develop the controller during the control design.

Function Approximator:

- Neural Network is used as a Function Approximator.
- An approximation may be required for some unknown function.

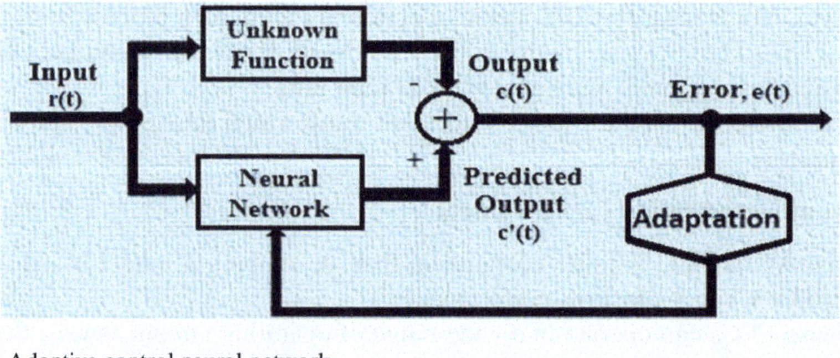

Fig. 9 Adaptive control neural network

- If the identical input is given to both systems, the network parameters should be changed to ensure that the system responds in the same way as the unknown function.

8.4 Hybrid Control Systems

A system made up of two dissimilar components is generally referred to as a hybrid. A hybrid system is one that combines digital and analog components, and it involves the interaction of event-driven and time-driven dynamics (constantly driven by commands and signals). The signals produced by such a system are a combination of continuous and discrete, taking values in a discrete set (like {a,b,c,...}) and a continuum (like the real numbers R). A hybrid control system can be represented as either the plant or the controller, separately or in combination [25].

Discrete planning algorithms and continuous processes combine to form hybrid control systems. Advanced control algorithms in computer control systems can be applied to hybrid control systems. This framework is helpful for simulating systems in a wide range of engineering applications: in mechanical systems, switches can be used to open and close or diodes to turn on and off continuous phenomena like capacitor charging; in chemical process control, pumps and valves can be used to control the continuous evolution of chemical reactions; and in embedded computation systems, a digital computer interfaces with an environment that is primarily analog.

9 Implementation of Control Systems in Mechatronics

9.1 Actuators and Sensors

Mechatronic systems, which combine electrical and mechanical components, are referred to as "smart" systems due to the incorporation of control, actuator, and sensor systems. The measurement system monitors changes in physical parameters like temperature, pressure, displacement, and other factors and then transforms them to electrical values like electric voltage or current, which is an essential part of a mechatronics system. The sensor and the signal-conditioning circuit are the two primary elements of the measurement system. A device that shows, documents, or regulates the parameter being measured causes the use of the output of the measurement system.

Actuators are machines that generate motion or action. It transforms energy from a source of energy—primarily electrical signals, air, or fluids—into motion. An actuator is an element via which a control system communicates with its environment. The mechanical output of an actuator is typically linear displacement or velocity. A few microns to a few meters can be the range of an actuator. Figure 10 shows a functional diagram of the actuator.

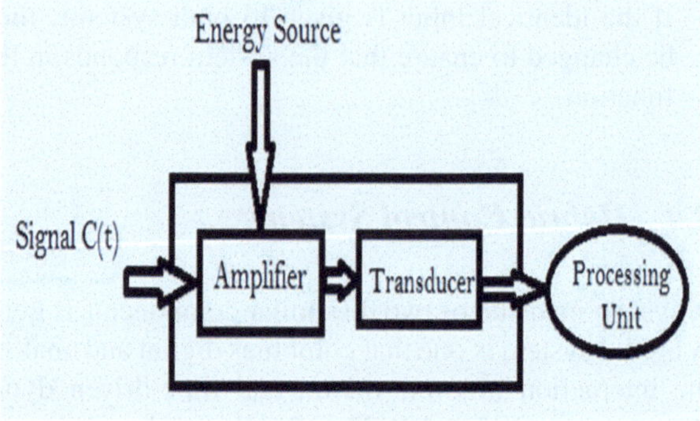

Fig. 10 Functional Diagram of an Actuator

Actuator Functional Diagrams:

- The (low-power) control signal is transformed into a high-power signal via the amplifier.
- The transducer creates work by converting the amplified control signal's energy.

Main Characteristics of Actuators:

The selection criteria of an actuator are influenced by several elements both in terms of functionality and performance. Listed below are many of these selection criteria:

- **Motion Conditions:**

The type of motion required, namely whether linear or rotary motion, or both, is one of the main criteria for choosing an actuation system. Moreover, rotational angles or stroke lengths must be specified. It might also be essential to determine the routing requirements, or the path that the actuator must follow.

- **Speed:**

Applications may have different needs for velocity. For instance, pick-and-place systems typically have high velocity requirements. However, for safety considerations, a robot with limbs may have lower velocity requirements. Another functional component is the demand for angular velocity. Velocity has an impact on the characteristics of acceleration and deceleration. High acceleration and deceleration requirements are usually required in pick-and-place systems because the mechanism must travel from minimum to maximum speed and vice versa in the smallest amount of time.

- **Power of Operations:**

The actuator's power development must be adequate for the purpose, that is, it must be able to generate the necessary force or torque. For instance, an actuator on a camera used for focusing will only need a little amount of power, yet a transportation system may need to move big loads. Power generation may need to account for variable loading, in which the actuator's load may fluctuate significantly while it is operating.

- **Resolution, Precision, and Consistency:**

Precision of an actuation system is determined by the aforementioned characteristics. The smallest controllable change in output that the actuator is capable of producing is known as resolution. A pick-and-place robot that locates components in a printed circuit board with a minimum separation of 2 mm is an example of a robot with resolution requirements. In this scenario, the robot's positioning resolution cannot exceed 2 mm.

Accuracy determines how close an actuator is predicted to place itself to the intended target by measuring the variation from the expected and obtained responses that the actuator achieves. While external forces like sliding friction may also have an impact, internal actuator factors like the looseness of mechanical connections and component selection are typically responsible for accuracy.

The actuator's repeatability is its capacity to return to the same location on multiple occasions. An actuation system's repeatability is often impacted by the stiffness of its components, particularly when the system loading varies. Additionally, system backlash and actuator hysteresis have a significant impact on repeatability.

- **Dynamic Action and Reactivity:**

Combination of the actuator's acceleration and velocity characteristics determines part of its dynamic response to the demand placed upon it. The dynamic response is a key factor in determining the actuator's dynamic properties and is heavily reliant on the efficiency of feedback from the device's sensors and control system. An actuator's dynamic behavior can be analyzed using first-, second-, or higher-order models that are appropriate for the task. Depending on the actuator's intended use, these system models can then be appropriately analyzed using frequency response analysis or step response analysis.

- **Compliance:**

A component moves in response to a force applied to it, which is known as compliance. Compliance can be characterized as the actuation system's sponginess/softness (high compliance) or rigidity (low compliance). Various actuation systems may have various compliance needs. For example, in order to provide the necessary machine tool precision during machining, actuation systems in machine tools would need to be highly stiff and, consequently, low compliance. However, great compliance would be necessary for a positioning system, such as the one used to position a tool in a tool holder, in order to facilitate the tool's easy insertion. In a measurement system, a sensor identifies changes in the physical parameters, but often the variable that is measured does not appear in the desired shape for the other system elements. Frequency-to-voltage conversion, sensor output amplification, analog-to-digital conversion (ADC), compensation, and other procedures are illustrations of signal-conditioning components. The signal-conditioning interface, an integral part of the measuring system, converts the measured signal into the required analog or digital electrical signal. Accurate linearized sensor data are also necessary for the operation of automated production systems and industrial processes, even in situations where a large number of sensors exhibit nonlinear properties. Accurate dynamic variable measurement is essential to the proper operation of the actuators in mechatronic systems. An electrical component on a printed circuit board could get misplaced as a result of a robot arm's rotational position being measured incorrectly. Because the temperature sensors installed inside the motor windings produce a nonlinear voltage output, the motor may shut down if the temperature reading is lower than the actual value.

Sensors now include integrated signal-conditioning circuits thanks to emerging technology in measurement systems. With the increased use of microcontrollers and microprocessors in sensing technologies, it has become more common to employ the signal-processing capabilities built into the sensors. The outputs provide precise and linear signals as a result. That's why selecting the appropriate linearization strategy is essential. Fast reaction and low sensor cost additionally require reducing the processor's response time and usage of memory. Research topics vary widely that involve designing hardware and/or software-based linearization and calibration approaches. Whether a mechatronics system is continuous or discrete, the capacity to detect its physical properties is essential to it.

9.2 Real-Time Control Systems

A system that could generate an appropriate answer in a set amount of time is said to be operating in real time. The process output would be impacted if the controller response was exceeded beyond these time constraints. In essence, real time refers to a value that is measured using a real clock. A real-time system receives inputs from the plant and sends the plant the appropriate control signal at the appropriate moment for the plant to operate. If a system uses the expression of time to describe its behavior, it can be referred to as a real-time system. Any signal is essentially processed by the real-time system, and it needs to react to the given input in a predetermined amount of time. It is not possible to classify this kind of system as a fast operating system.

Types of Real-Time System:

- **Clock-based system**

A time constant, which measures how long it takes the plant to react to changes in the input or load, is measured in this kind of control system. In the case of a chemical process, the time constant might be measured for specific hours. The time constant of the process that needs to be regulated determines the necessary sample rate in feedback control. If the processing time constant is small, the sample rate will be substantially faster. An additional real-time clock will be attached to the computer system, and it will send a signal to the computer system at the appropriate time. In a process plant, this kind of system is utilized to carry out cyclic or periodic duties.

- **Event-based task**

Because actions in this kind of system are based on events, the controller would only take action in response to an event. As a result, this kind of control system doesn't operate on time intervals, and one of its key needs is that it reacts to events promptly. When the water level in the tank reaches its maximum, this system is utilized to close a valve.

- **Soft real-time system**

This system's performance will deteriorate if it is unable to generate the needed result at the appropriate moment. Therefore, if part of the soft real-time system fails to meet the deadline, it won't have an impact on the process. Therefore, there won't be any issues if overall performance is strong, but this isn't the case with the hard real-time system. Robotic assembly lines in manufacturing are an example of a soft real-time system. Therefore, if the deadline is missed in this production line, the parts will not be assembled correctly. Production will therefore continue if the damaged part is not too expensive and is found by quality control on an occasional basis. This kind of system will respond slowly if the load is really high.

- **Hard real-time system**

The operation of this system will be incorrect if the results are not produced within the specified time frame, as it is necessary for this system to do the necessary action on every single occasion. The entire process would be impacted if the hard real-time system's reaction time was off. The hard real-time system's peak load performance needs to be predictable and shouldn't go over the allotted time. This system needs to always be in sync with the condition of the environment. The hot air blower system's temperature control loop serves as an illustration of a hard real-time control system.

10 Applications of Control Systems in Mechatronics

10.1 Robotics and Automation

Automation is the term used to describe the automatic control of industrial and other operations, as noted before, as opposed to manual control. Not only has automation been essential to modern industrial processes, but it has also been to traffic, robotics, and automobile systems. The steel, chemical, electric power, and automotive industries are typical instances of industries that automate. Self-acting machinery and devices with integrated control systems, frequently set up in intricate industrial processes or systems, are what define automation.

Automation includes measuring, controlling, monitoring, supervising, and other activities. It is typically organized in many hierarchically structured layers and managed by a supervisory control system. Examples of automated processes include tooling, handling, and assembly. Automation makes it possible to achieve the best possible product quality, boosts output, and frees up human labor for a variety of repetitive, routine, and boring tasks. Automation is also capable of tasks that are well above human physical capabilities, including precisely locating a huge radio telescope at a predefined spot.

Robots can be controlled using a variety of techniques, the most popular of which are PID (Proportional-Integral-Derivative) control systems. The best illustration is how it corrects a control function with accuracy and responsiveness; cruise control is one such example. PID is widely used since there is a growing need for more efficient controllers due to the increased application of robots. Using linear quadratic regulators, or LQRs, is another well-known method that provides "optimally controlled feedback gains to enable the closed-loop stable and high-performance design of systems" (source). Another well-known technology is radio frequency identification, or RFID, which is widely used to enable wireless access to control systems.

10.2 Automotive Systems

10.2.1 Engine Management System (EMS)

An electronic control unit monitors engine speed performance to regulate engine operation. It also estimates the precise fuel needed for starting an engine. Cars are equipped with a variety of subsystems that can be used: MAP Sensor, Knock Sensor, Engine Speed Sensor, Engine Oil Sensor, Crankshaft Sensor, Camshaft Sensor, Multi-point Fuel Injection, Air Flow Sensor, Air Temperature Sensor, Throttle Position Sensor (TPS), Oxygen Sensor, Idle Air Control Valve (IACV), Electronic Control System, Electronic Control Unit, and Electronic Fuel Injection System.

- Electronic Fuel Injection System: Control when to ignite the spark plugs and how much fuel to inject into the engine. Use the least amount of fuel to produce the greatest horsepower.
- The Electronic Control Unit (ECU) aids in the analysis of fuel injection at the proper time and pace.

10.3 Aerospace Systems

Control systems are employed in the production, upkeep, and repair of spacecraft, as well as private and military automobiles and aircraft. They precisely lift and position airplane components, lowering the possibility of human error. Control systems are made to meet the aerospace industry's growing need for accuracy and productivity. These systems need to be flexible because the machinery in this industry is always evolving and inventing. Control systems that provide personalized feedback and automation functionalities are therefore essential. Faster than ever before, safer items are available for use thanks to more accurate and efficient systems. Furthermore, one system may perform functions for a variety of aircraft thanks to adaptable control systems. As a result, less expensive specialized equipment is required. Engineers can use present equipment to plan for future model changes thanks to technological advancements in control systems.

Large or intricate pieces define the assembly and manufacturing process of aircraft. Control systems are used by manufacturers to precisely position these parts. The lifting components of the systems need to be strong and low-maintenance, able to support large weights. Aside from that, components of an aircraft can be specially designed or connected to facilitate fuel-efficient flight. These factors must be taken into consideration since they can contain fragile, light components.

Aircraft are manufactured, repaired, and maintained using control systems. Because safety comes first, there are strict maintenance requirements. Automated control methods greatly increase the efficiency of operating in tight or problematic spaces, like airplane wings. Deploying control systems in high volume, repeated refurbishing, assembly, or repair activities is ideal.

10.4 Biomedical Systems

The development and execution of systems with the ability to control and modify the behavior of other systems, such as biological, chemical, and physical processes, is the focus of the field of control engineering. To accomplish desired outputs and performance, control engineers employ mathematical models, algorithms, sensors, actuators, and feedback systems. Applications of control engineering to human health and medicine include biomedical systems, which include biosensors, drug delivery, artificial organs, rehabilitation, and diagnostics. The field of biomedical engineering is leading the way in revolutionary discoveries that are changing the face of healthcare. This multidisciplinary area combines biological and medical sciences with engineering principles to create ground-breaking solutions for medical problems and improve patient care. The discovery of wearable medical sensors and gadgets is one of the major new topics in biomedical engineering. These wearable or integrated into clothes gadgets continuously assess health indicators, monitor vital signs, and give real-time feedback to patients and healthcare providers. These technological advancements provide proactive and individualized healthcare solutions, ranging from biosensors that analyze sweat for biomarkers to smart watches that can monitor an ECG.

A further trend that is transforming healthcare is the quick development of 3D bioprinting and personalized medication. By utilizing developments in tissue engineering, genomics, and molecular biology, biomedical engineers are able to customize therapies and treatments according to a patient's genetic composition. Furthermore, the accuracy with which tissues, organs, and implants may be created thanks to 3D bioprinting technology that has the potential to transform regenerative medicine and organ transplantation.

The integration of artificial intelligence (AI) and machine learning is a significant advancement in biomedical engineering. Algorithms using AI evaluate vast amounts of medical data that help in discovering of new medications, the planning of treatments, and the detection of diseases. In the end, these technologies improve patient outcomes by enabling quicker diagnosis, more precise predictions, and customized treatment regimens.

Moreover, nanotechnology has changed the game in the field of biomedical engineering. Targeted medication delivery methods, accurate imaging methods, and even the possibility of early illness detection at the cellular or molecular level are all made possible by nanoscale particles and devices. These developments have enormous potential for less invasive and more successful medical procedures.

Even with these encouraging developments, widespread adoption still faces obstacles like obtaining governmental permissions, protecting user privacy, and ensuring accessibility. Yet, the combined efforts of regulatory agencies, medical professionals, and biological engineers keep pushing these advancements forward, transforming healthcare and opening the door to a healthier future.

10.5 Conclusion

More affordable, faultless, and superior products are needed to meet the expectations of the modern industry. As conventional human-based manufacturing processes are unable to meet this criterion, automation and industrial applications cannot currently be separated. Control and automation engineers with knowledge and experience in process control, control system design, and problem solving are typically required. In its broadest sense, control and automation engineering seeks to find a solution of coercing a system into the desired state by the use of appropriate hardware and software. After the Industrial Revolution, people searched for alternative sources of income due to a number of problems caused by the rapid increase in output and the rapid advancement of technology. Furthermore, the expanding building sector and technological developments have elevated intelligent buildings to a significant role in our daily lives. The development of an automatic pilot to allow unmanned ground and aerial vehicles to follow the specified course is an example of an application of control and automation engineering. Although their success will depend on how far the aerospace and aviation industries advance, satellite launches have begun. Self-reliance has become so prevalent. It has started to make way for these developments in other control domains and the aerospace and aviation industries. Medical and modern technology developments are beginning to form a single framework and create new operating conditions. Bionic arms and surgical robots are two well-known examples of these developments. Mechanical systems and the feedback mechanism of the human body are very similar since they are both feedback control systems.

Technology has revolutionized control systems, making them smarter, more efficient, and innovative in the process. Control systems have advanced beyond their previous constraints to provide unmatched precision and intelligent features. The development of mechatronics, which ushers in a new era of automation, machine learning, and networked intelligence, has sculpted a world of new possibilities. The most recent trends and projections provide us with a glimpse into the future of control systems. Exploring emerging technologies and their impact on related businesses, a projection for the future can be illustrated. Contemporary innovations and major infrastructures, as well as the incorporation of state-of-the-art technologies into control systems, pose a number of ethical concerns. The majority of these concerns center on significant issues such as ensuring privacy and security. They also raise the possibility that automation will someday take the role of human operators. Consequently, moderation and cautious consideration are needed. Technology-integrated control systems raise concerns regarding algorithmic bias, privacy, and accountability. These issues must be resolved in order to guarantee the moral and proper usage of emerging technologies. For control systems to endure, responsible design and management are essential. They ensure that risks are kept to a minimum and moral considerations are made for the benefit of society at large. Emerging technology has the potential to bring about a revolutionary breakthrough in control engineering. Continuous learning, adaptation, and creativity are required to reach this potential and progress the sector.

References:

1. Salem FA, Mahfouz AA (2013) A proposed approach to mechatronics design and implementation education-oriented methodology. Innov Syst Des Eng 4
2. Salem FA, Mahfouz AA (2014). Mechatronics design and implementation education-oriented methodology; A proposed approach. J Multidiscip Eng Sci Technol 1(3)
3. Salem FA (2015) The role of control system/algorithm subsystems in mechatronics systems design. J Multidiscip Eng Sci Technol 2:2717–2732
4. Avram M, Bucşan C, Duminică D (2011) The control subsystem of a mechatronic system. RomIan Rev Precis Mech 40:168
5. Spiegel MR (1965) Schaum's outline of theory and problems of Laplace transforms. (No Title)
6. Zwart H (2004) Transfer functions for infinite-dimensional systems. Syst Control Lett 52(3–4):247–255
7. Svensson A, Schön TB (2017) A flexible state–space model for learning nonlinear dynamical systems. Automatica 80:189–199
8. Rosenvasser YN, Polyakov EY, Lampe BP (1999) Application of Laplace transformation for digital redesign of continuous control systems. IEEE Trans Autom Control 44(4):883–886
9. Kiran CP, Clement S, Agrawal VP (2012) Quality modelling and analysis of a mechatronic system. Int J Ind Syst Eng 12(1):1–28
10. Tornambe A (1999) Modeling and control of impact in mechanical systems: Theory and experimental results. IEEE Trans Autom Control 44(2):294–309
11. Lai CS, Locatelli G, Pimm A, Wu X, Lai LL (2021) A review on long-term electrical power system modeling with energy storage. J Clean Prod 280:124298
12. Janot A, Young PC, Gautier M (2017) Identification and control of electro-mechanical systems using state-dependent parameter estimation. Int J Control 90(4):643–660
13. Gabano JD, Poinot T (2011) Fractional modelling and identification of thermal systems. Signal Process 91(3):531–541

14. Hua MD, Hamel T, Morin P, Samson C (2013) Introduction to feedback control of underactuated VTOLvehicles: A review of basic control design ideas and principles. IEEE Control Syst Mag 33(1):61–75
15. Zhang F, Sun Y, Cheng L, Li X, Chow JH, Zhao W (2014) Measurement and modeling of delays in wide-area closed-loop control systems. IEEE Trans Power Syst 30(5):2426–2433
16. Caldwell J, Wang W, Zandstra PW (2015) Proportional-integral-derivative (PID) control of secreted factors for blood stem cell culture. PLoS ONE 10(9):e0137392
17. Walsh GC, Ye H, Bushnell LG (2002) Stability analysis of networked control systems. IEEE Trans Control Syst Technol 10(3):438–446
18. Choudhury DR (2005) *Modern control engineering*. PHI Learning Pvt. Ltd
19. Calafiore GC, Campi MC (2006) The scenario approach to robust control design. IEEE Trans Autom Control 51(5):742–753
20. Schaal S, Mohajerian P, Ijspeert A (2007) Dynamics systems vs. optimal control—a unifying view. Prog Brain Res 165:425–445
21. Dumont GA, Huzmezan M (2002) Concepts, methods and techniques in adaptive control. In Proceedings of the 2002 American control conference (IEEE Cat. No. CH37301), vol 2, IEEE, pp. 1137–1150
22. Farhan A, Wei CL, Ahmed MT (2018) A qualitative overview of fuzzy logic in ECG arrhythmia classification. Int J Eng Work 5(11):232–239
23. Bai Y, Wang D (2006) Fundamentals of fuzzy logic control—fuzzy sets, fuzzy rules and defuzzifications. Adv Fuzzy Log Technol Ind Appl 17–36
24. Gomi H, Kawato M (1993) Neural network control for a closed-loop system using feedback-error-learning. Neural Netw 6(7):933–946
25. Lennartson B, Tittus M, Egardt B, Pettersson S (1996) Hybrid systems in process control. IEEE Control Syst Mag 16(5):45–56

Robotics in Mechatronics Engineering

Md. Tarek Ur Rahman Erin, Ahmed Farhan, Protik Barua, Naheen Ibn Akbar, Rezwan Us Saleheen, and Md Sadatuzzaman Saagoto

Abstract This chapter, "Robotics in Mechatronics Engineering," provides an overview of the area of robotics, including its essential components, the course of history, and integration with mechatronics. This chapter investigates robotic kinematics and dynamics, including topics like location, velocity, motion control, and forward and inverse kinematics. It categorizes robots based on their applications, focusing on industrial, service, and collaborative robots, each with distinct characteristics and use cases. Furthermore, the chapter stresses the importance of robot vision and perception by discussing how robots utilize sensors and cameras to see their surroundings, recognize objects, and apply spatial intelligence, all of which contribute to their autonomy. This comprehensive outline establishes a solid basis for robotics within mechatronics and its numerous implications for various sectors.

Keywords Robots · Robotics · Mechatronics · Sensors · Automation · Controller

1 Introduction

The advancement of contemporary technology has gained a different dimension with robotics and mechatronics engineering in recent years. Mechanical, electrical, and computer engineering are combined in a way that works well to provide the theoretical foundation for robotics and mechatronics engineering. This area of engineering is quite young. The word "mechatronics" emphasizes how important it is to combine mechanical and electrical systems into a particular device. Mechatronics is the field in which engineering science, applied science, and engineering principles come together to facilitate the design, fabrication, and use of devices. As was previously said, robotic technologies may be thought of as a subset of mechatronics. The

Md. T. U. R. Erin
Department of Mechanical Engineering, Textile Engineering College, Chattogram, Bangladesh

A. Farhan · P. Barua (✉) · N. I. Akbar · R. U. Saleheen · M. S. Saagoto
Department of Mechatronics Engineering, World University of Bangladesh, Dhaka, Bangladesh
e-mail: protik.barua@mte.wub.edu.bd

M. M. Rahman et al. (eds.), *Mechatronics*, Emerging Trends in Mechatronics,
https://doi.org/10.1007/978-981-97-7117-2_4

procedure begins with the device's design and ends with the user-built equipment operating without a hitch [1].

1.1 Overview of Robotics

Engineering, science, and technology come together in robotics to create the machine known as a robot. The field of robotics involves the development and use of robotic systems. Robots are mostly used to replace people in a variety of activities. Robot design, building, and operation are the main topics of robotics, together with computer systems for control, feedback, and information processing. It is a machine that has been programmed to carry out the assigned duties and collect data from its environment. It has sensors to perceive its environment and is driven by a central CPU to regulate its motions [2].

1.2 Mechatronics: Integration of Mechanical, Electrical, and Computer Engineering

Mechatronics was first used by the Japanese scientists in the late 1970s to refer to the approach they used while designing the electromechanical systems' subsystems. Mechatronics is the design of goods and manufacturing processes that combine the synergistic elements of electrical, mechanical, system, and control engineering. The integrated and multidisciplinary approach to engineering design is becoming more and more common to design automobiles, machine tools, robotics, cameras, and other devices. We need to incorporate mechanical, electrical, electronic, and control engineering from the outset of the design process in order to create systems that are more affordable, dependable, and adaptable. Systems for mechatronics are developing with an emphasis on efficiency and automation. These systems start to behave more and more like robots as they become more independent. Common illustrations of mechatronics projects include robot controllers, biped walking robots, robotic hands, animatronic hands, and manufacturing plant automation.

Mechatronics brings the areas of technology together involving sensors, actuators and drives, measurement systems, and microprocessor systems, as well as the control system behavior and analysis of system [3]. Figure 1 illustrates these basic elements of a mechatronics system.

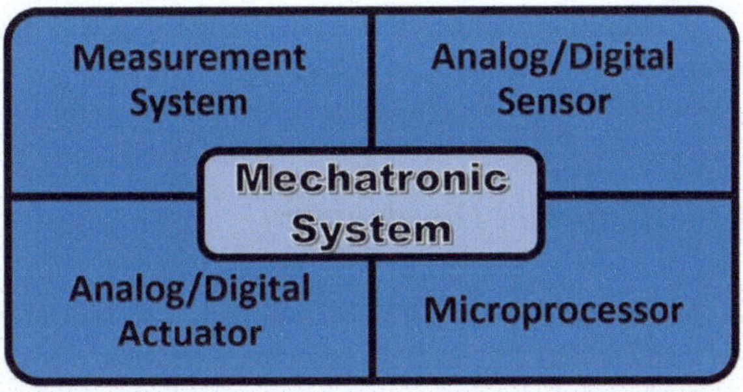

Fig. 1 The major elements of a mechatronic system

1.3 Significance of Robotics in Mechatronics

Robotics and mechatronics are two fields that have a lot in common. Both involve the design and construction of machines that interact with their environment in some way. Both necessitate a profound comprehension of the functioning of these machines, enabling efficient and effective design. Robotics and mechatronics are the points at which mechanical and electrical engineering come together, using computer-controlled systems to make devices smarter and more efficient. Robotic systems assemble mechanical, electrical, and software components to accomplish the desired function.

Robotics has led to the creation of a new field of study called mechatronics, which focuses on the interaction between robots and humans. Today, the manufacturing industry most commonly uses mechatronics, employing robots to perform tasks that would otherwise require significant human time and effort [4].

2 Fundamentals of Robotics

The need for automation with industrial robots is still growing in spite of a lack of skilled labor and the shifting of manufacturing to high-wage industrial nations. For the correct design of robot systems, it is essential to understand the many types of kinematic robots, their benefits and drawbacks, the unique motion behavior of industrial robots and how to mathematically describe them, the parts of automation systems, and, of course, safety considerations.

Understanding robotics begins with an understanding of the many types of kinematic robots, their characteristics, and their uses. It focuses on the scientific description of robotic motions as utilized for robot simulation and control, the calculation of rotations, and the application of Denavit-Hartenberg rules to define kinematic chains. It also addresses the fundamentals of controlling motion and path planning, robot programming, and the systematic design of general handling systems which includes interactive, teach-in, and automatic offline programming. It also includes safety equipment, characteristics of safe robot cell design, robot hardware, and the precision of robot-based actions [5].

2.1 Historical Development

Modern culture is driven by technological breakthroughs in fields such as robots, electronics, and software. Every day, technology advances and transforms the way we work and do business. Autonomous robotics and artificial intelligence (AI) technologies are arguably at the forefront of contemporary technological innovation. Undoubtedly, mobile, self-governing robots will be important in the future.

The business of commercial robot has progressed significantly in only fifty years. "Cobots" or collaborative robots can operate alongside human beings. These robots are coined as a result of the progression of robotics in the commercial field. The question of whether autonomous robots will eventually replace humans in jobs or work alongside them is still up for debate [6].

We must look back to the 1940s and 1950s, when robots first appeared, in order to comprehend the effects, they have on various businesses. British neurophysiologist W. Grey Walter is the pioneer of this industrial robot who is frequently credited with creating the two robots, Elmer and Elsie, in the late 1940s. The Grey Walter's Tortoises demonstrated basic autonomous behaviors that were modeled after biological systems. The robots could traverse their environment and even react to stimuli using simple sensors, mimicking human behavior. A brief historical advancement of robot is illustrated in Table 1.

Table 1 Evolution of robots and robotics

Time	Significance	Name of the robot	Inventor
First century A.D. and earlier	Pneumatica and Automata by Heron of Alexandria contain descriptions of over a hundred machineries and automata, such as a steam-powered engine, a coin-operated machine, a wind organ, and a fire engine	None	Heron of Alexandria, Ctesibius, Philo of Byzantium, and others
1206	The first robots with humanoid programming	Four robotic musicians on a boat	Al-Jazari
1495	Humanoid robotic designs	Mechanical knight	Leonardo da Vinci
1738	A mechanical duck with the ability to fly, eat, and urinate	Digesting Duck	Jacques de Vaucanson

(continued)

Table 1 (continued)

Time	Significance	Name of the robot	Inventor
1800s	Japanese mechanical toys with painted surfaces, arrow shooters, and tea services	Karakuri toys	Tanaka Hisashige
1921	The first imagined automata referred to as "robots" first appear in the drama R. U. R	Rossum's Universal Robots	Czech writer Karel Čapek
1930s	Robotic humanoid displayed at the 1939 and 1940	Elektro	Westinghouse Electric Corporation
1949	Basic robots with self-navigation behaviors	Elsie and Elmer	William Grey Walter
1956	Based on George Devol's inventions, the Unimation firm, created by Joseph Engelberger and George Devol, produced the first commercial robot	Unimate	George Devol
1961	Commercial robot installation	Unimate	George Devol
1963	Robot for palletizing	Palletizer	Fuji Yusoki Kogyo
1973	The first six electro-mechanically powered axes of an industrial robot	Famulus	KUKA Robot Group
1975	Universal manipulation arm	PUMA	Victor Scheinman

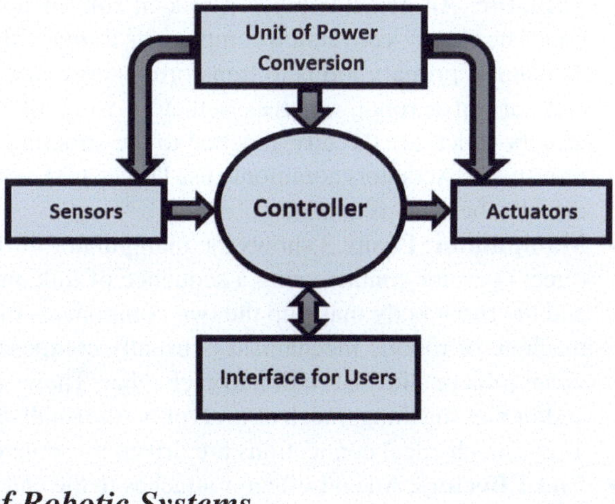

Fig. 2 Key components of robotics

2.2 *Key Components of Robotic Systems*

A robot is usually an electromechanical system consisting of some components like power supplies, sensors, actuators, controllers, and some other components [7]. Figure 2 illustrates the basic components of a robotic system.

The major components are identified by their respective names.

- **Power supplies**: The primary function of a power supply unit is to provide power to an electrical load and to transform current of a source into the proper frequency and voltage. As a result, they are occasionally denoted as electric power converters.
- **Feedback devices/Sensors**: A sensing unit is an apparatus that offers data on the external environment. These gadgets belong to a class that is used to watch over a process or operation and then confirm that it actually took place. A sensor is an apparatus that measures and converts a physical quantity into a signal so that an instrument or observer can understand. Returning to the analogy of a human body, sensors are the organs that provide information to the brain (controller). To detect the locations of different joints and linkages, a robot might attach a feedback device and provide data to the controller. The robot's arm may operate basic limit switches or position-measuring tools like encoders, resolvers, tachometers, and/or potentiometers. The feedback data is either digital or analog based on the devices.
- **Controllers**: The robot controller is a software and hardware-based device that is connected to the robot and serves as its brain. The robot part that manages the movements of the entire mechanical system is called the controller. It also uses a range of sensors to gather information from its immediate surroundings. A microprocessor at the heart of the robot's controller connects to the input/output and monitoring devices. The controller issues a command that activates the motion control mechanism, comprising various controllers, actuators, and amplifiers. The controller unit receives the signal from the sensor and drives the actuators. Consider the Arduino Uno, a microcontroller board that relies on the ATmega328.
- **Actuator**: An actuator plays a crucial role in powering a robot's mechanical unit. The energy-conversion components found within robots are called actuators. Actuators' primary job is to transform energy into motion. Sensors are the eyes and ears of a robot, whereas actuators work like muscles. Actuators are tiny motors which are directly attached to the structure of the machine that facilitate movement. Actuators commonly use DC-geared motors. Servo and stepper motors can also be used as actuators.
- **Manipulator**: Figure 3 shows the manipulator of the parts in the robots without direct operator contact using a sequence of link and joint connections. The arm and the robot body make up the two components of the robot manipulator [8]. A machine or robotic mechanism is usually composed of many segments that are either joint or slide in relation to each other. These segments are used for gripping and/or moving things, such as tools or parts, usually in several degrees of freedom. These mechanical connections are driven by actuators.
- **End-Effectors**: An end-effector attaches to the end of a robotic arm to operate. It moves or orients the process or product. Specialized tasks like welding, measuring, marking, drilling, cutting, painting, and cleaning utilize it [9].

Fig. 3 Different types of workspaces

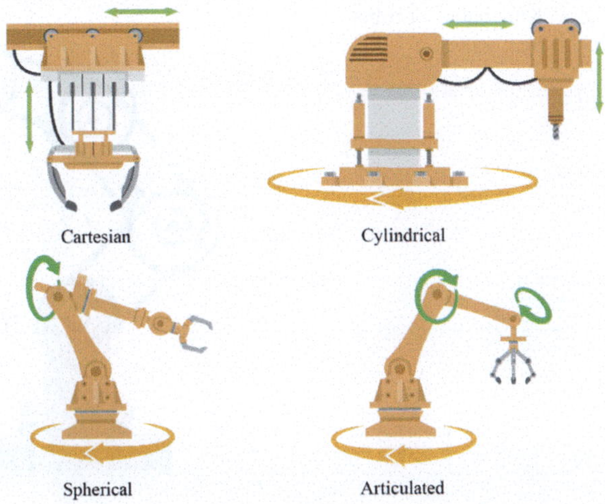

2.3 Robot Joints

Because they enable the connections to move in a variety of ways, the robot joints are the most crucial component. Joints come in five main categories.

- **Linear Joint:** A linear joint is represented by the symbol L. This kind of joint has the ability to move both slidingly and translationally.
- **Orthogonal Joint:** Symbol "O" is used to represent the orthogonal joint. It has certain similarities with the linear joint. The input and output connections will move at right angles, which is the sole change.
- **Rotational Joint:** Symbol R can also be used to symbolize a rotating joint that can rotate along the axis that runs parallel to the axes of the arms.
- **Twisting Joint:** It is known as a twisting joint, joint T. The movement of this joint between the input and output links is twisting. The rotation axis and the output link's axis are vertically aligned throughout this operation. The output link rotates with the relation to the input link.
- **Revolving Joint:** The V joint is the popular name for the revolving joint. The input link is parallel to the rotating axes and the output link axis is perpendicular to the rotational axis. The output link rotates around the input link like a twisting joint [10].

2.4 Classification of Robots in Mechatronics Context

There are many different sizes and forms of robots as shown in Fig. 4. They can perform a variety of arm manipulations and motion systems. Most commercially

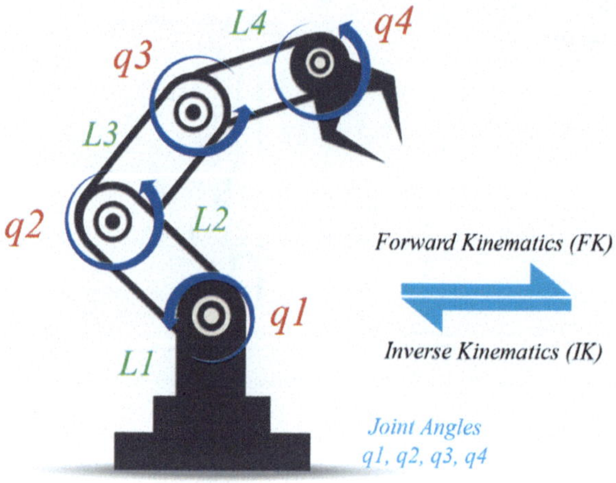

Fig. 4 Kinematics of robot

available industrial robots have five basic configurations based on their physical characteristics. There are, nonetheless, a number of other kinds of robot setups.

There are various types of industrial robots, as outlined below.

2.4.1 Cartesian/Rectangular/Gantry Configuration (PPP)

This configuration uses three perpendicular slides for the construction of the x, y, and z axis. The robot can operate within a rectangular work envelope by moving three slides relative to one another. They are also known as Rectangular or Cartesian configuration robots.

2.4.2 Cylindrical Configuration (RPP)

These robots make use of a slide-moving vertical column. This robot may travel radially with respect to the column by fastening an arm to the slide. The robot may retrieve a cylindrical work envelope by rotting the column.

2.4.3 Spherical/ Polar Configuration (RRP)

The robot utilizes a telescopic arm with the ability to raise or lower itself around a horizontal pivot point. A rotating base mounts the pivot point, enabling the robot to move vertically. The robot can move its arm within a spherical enclosure.

2.4.4 Articulated/ Revolute/ Jointed Arm/ Anthropomorphic Configuration (RRR)

It is made up of two straight halves, the elbow and shoulder joints which revolve along horizontal axes that represent the human upper and forearm. The shape of the work envelope is not uniform.

2.4.5 SCARA Configuration (RRP)

This particular kind of the jointed arm robot rotates the elbow and shoulder joints along vertical axes rather than horizontal ones. Its cylindrical work envelope, which is far greater than that of any other arrangement, offers significant vertical stiffness for a number of crucial jobs [11]. Robots have several types, like Cartesian, spherical, articulated, and cylindrical [12].

3 Work Envelop/Work Space

The region encompassing all points in the surrounding space that the robot arm can reach is referred to as the work envelope (also called the robot reach or work volume). It is one of the most crucial factors to take into account when choosing a good robot since the application space must remain within the chosen robot's reach. The collection of places that a robot manipulator's end-effector can reach is known as its workspace. Stated differently, a robot's workspace is the area where its mechanism operates. Workspaces have different categories, for example, cylindrical, Cartesian, articulated, and spherical [13]. These categories of workspaces are shown in Fig. 3.

- **Cartesian Work Envelope**: This is a rectangular prism that represents a Cartesian configuration envelope. Its work envelope is devoid of dead zones, allowing the robot utilizing it to freely operate the highest payload it is capable of handling without experiencing any strain. This kind of work envelope is present in gantry or Cartesian robots.
- **Cylindrical Work Envelope**: This arrangement, as its name implies, is cylindrical in shape and has a hollow center, which significantly restricts the robotic arm's range of motion. This arrangement leaves a sizable dead zone both inside and outside the robot frame. This kind of envelope is most appropriate for short, simple operations requiring little movement performed by robots.
- **Spherical/ Polar Work Envelope**: The volume situated between two half spheres is swept up by this arrangement. The robotic arm's maximum angular rotations in both the vertical and horizontal axes are restricted by this design. Unfortunately, these physical constraints result in large, noticeable dead zones above and below the concerned robot.

- **Articulated/ Revolute Work Envelope**: Usually, this unusual arrangement has an envelope larger than the available floor area. In essence, the robotic arm is supported by a large, open work envelope that may be either completely closed or partially opened. In terms of work envelope arrangements, it is among the most intricate.
- **SCARA Work Envelope**: This arrangement is unique to SCARA robots and usually resembles a kidney or heart. Its distinctive hollow circle, which runs directly through the center, facilitates simpler lateral motions and offers a very wide horizontal plane of operation. Vertical motions are, however, limited since there is very little room remaining for upward motion [14].

Dead Zone: Often, the robot arm cannot reach certain areas within the working envelope. These areas are known as dead zones.

4 Kinematics and Dynamics

The geometry and algebra of robot motion are studied in robotic kinematics. It explains the relationship between a robot's joints' position, orientation, velocity, and acceleration, as well as their links to each other and the environment. Typically, the robot kinematics have two types: forward and inverse. Given the joint angles, forward kinematics calculates the end-effector pose (the position and orientation of the robot's tool or sensor). Inverse kinematics computes the joint angles given the desired end-effector pose. Robot kinematics is essential for planning and executing robot motions, such as reaching, grasping, or manipulating objects.

Robot dynamics is the study of the forces and torques that affect robot motion. It describes how mass, inertia, friction, gravity, and external forces influence the robot's acceleration and energy. Robot dynamics can be classified into two categories: direct and inverse. Direct dynamics computes the joint torques given the joint angles and accelerations. Given the joint angles and torques, inverse dynamics computes the joint accelerations. Robot dynamics is important for designing and controlling robot systems, such as balancing, stabilizing, or optimizing performance [15].

4.1 Basics of Robot Movement Control/Trajectory of a Robot

Trajectory planning is critical for robotic applications and automation in general. The capacity to produce trajectories with provided characteristics is a critical element to ensuring substantial outcomes in terms of quality and convenience of accomplishing the required motion, especially at the high operating speeds needed in many applications. [16]. A trajectory is a robot's configuration as a function of time. There are four types of robot trajectory and motion control. They are mentioned below.

4.1.1 Point-To-Point Control Robot (PTP)

A time-optimal movement in three dimensions between two specified points is called a point-to-point movement (PTP movement). Robot axes are traveling from the present location (M1) to the target point (M2) in synchrony, which causes the end-effector's trajectory to curve. PTP motions are utilized for fast positioning, and then a targeted operation or controlled path motion starting at the target location is applied (M2). The PTP robot has the ability to move between locations. The locations are stored in the control memory [17].

4.1.2 Continuous-Path Control Robot (CP)

It is possible for the CP robot to go along the predetermined course. The robot may halt at any point in the controlled path with CP from a single control. Every point along the path needs to be explicitly stored in the robot's control memory. The most basic example of this kind of robot is one that moves in a straight line [18]. Several robots that use continuous pathways can also be trained to follow a smooth curve. In these cases, the controller unit stores a large number of discrete point locations along the path in memory (teach-in), and the programmer manually leads the robot arm along the desired path.

4.1.3 Controlled-Path Robot

Controlled-path robots are capable of producing very accurate pathways with varying geometry, including circles, straight lines, and interpolated curves. Almost every site on the given path may get high accuracy. Just the start and finish locations and the path definition function need to be kept in the robot's control memory [19].

4.1.4 Stop-To-Stop

The system is open-loop. The controller does not know the position or velocity. The on/off commands are stored by the controller as valve states. The final trip is also adjusted via a mechanical device.

4.2 Kinematic Analysis

The link between a robot's endpoint motion and joint motion is known as kinematics. It is a basic and traditional topic in robotic system that analyzes the link between the joint coordinates of a robot and its spatial arrangement. Figure 4 illustrates the kinematics of a robot. When calculating the location of a gripper in space, creating a mechanism to move a tool from one position to another position, or determining if a robot's motion will collide with an obstruction, kinematics can provide extremely precise results [20].

Characteristics including singularity manifolds, joint restriction, collision avoidance, and redundancy are all covered by robotic kinematics. For this reason, kinematic analysis is essential to a robot manipulator's positioning duty. The kinematics of the robot arms often divide into forward and inverse solutions.

4.2.1 Forward Kinematics

The end-effector placement of a robot is determined by computational processes. The processes employ combined sensors and mathematical algorithms to pinpoint its position. A manipulator is made up of serial links that are connected to one another from the base frame through the end-effector by prismatic or revolute joints. The location and orientation of the end-effector are determined by forward kinematics using the joint variables.

4.2.2 Inverse Kinematics

In inverse kinematics, the robot's motion to a target point is calculated using kinematic equations. For example, automated bin picking on a production line involves a robotic arm that moves precisely between bins and manufacturing units from one location to another. The grabbing end of the robot arm is called the end-effector. The robot configuration is a list of joint positions that do not deviate from any limitations the robot has and fall inside the position limits of the robot model.

4.3 Dynamic Analysis and Control

The link between the accelerations that a robot mechanism produces and the forces exerted on it is the focus of robot dynamics. Robot dynamics models the robot mechanism as a rigid-body system and applies rigid-body dynamics to robotics. Robot dynamics has two main problems, forward dynamics and inverse dynamics. Figure 5 represents these two problems of robot dynamics.

Forward dynamics is also referred to as "direct dynamics," or just "dynamics." It is mostly employed in simulation. Numerous applications of inverse dynamics exist, such as the online control of robot forces and movements, robot mechanism design, trajectory design, and optimization, and as a building block for some forward-dynamics methods [21].

Another type of dynamics is hybrid dynamics, which determines the unknown forces and accelerations given the forces at certain joints and the accelerations at others. This system can often flow and jump at the same time.

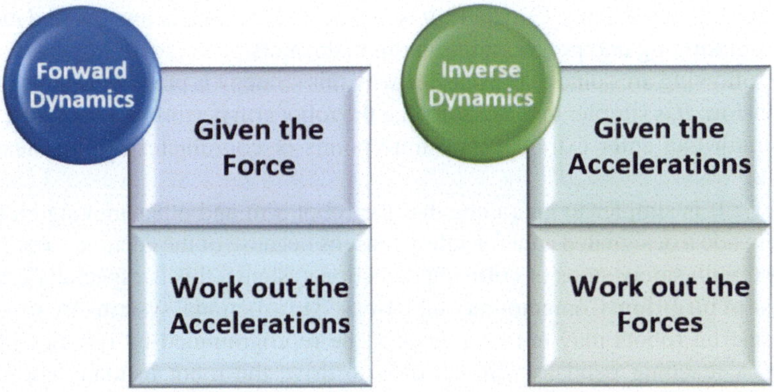

Fig. 5 Problems of robot dynamics

5 Programming and Control

Robot control is an essential technology that enables a robot to move precisely and adaptively. In order to ensure that the robot can do its mission and achieve the required transient and steady state, the formulation of the robot control issue entails defining the joint torques that the joint actuators will create. Robot controls are divided into two types based on the restrictions of the environment: motion control in unconstrained space and interaction control in restricted space.

5.1 *Programming Paradigms for Robotics*

Programming is a key aspect of any robotic system. It transmits vital information to an industrial robot, allowing it to function and perform specific tasks. Programming robots to perform their tasks can take various forms.

5.1.1 Online-Programming Method

- **Teach Method**:

With this technique, the robot is guided through a sequence of locations and their memory is retained by the use of a teaching necklace. Teaching in robotics refers to the act of directing and logging these locations, or coordinates, in space. Most contemporary industrial robots available today have a teach pendant that enables anybody, not only engineers, to operate the robot and train it to carry out certain tasks. Custom apps installed into a specialized tablet or touchscreen device constitute modern teaching pendants. Currently, it is the most user-friendly and favored method for reprogramming and programming industrial robots.

The following are some of the reasons why this strategy is preferred above others:

Precision: It is simpler to make sure that the robot arm performs as intended when the operator can enter extremely accurate points or coordinates into the teaching pendant.

Safety: It is simpler to guarantee that the robot arm and other moving elements remain inside a designated area for safety reasons because of the accuracy. For robots equipped with safety stops or collision detection systems, this is especially crucial.

Sense of intuition: Contemporary tablet-based instructional systems are so natural that industrial robots may not even need to be reprogrammed by robotics professionals. A teaching pendant with controls to drive the robot in many coordinate systems is used to manually drive the robot to the required places.

The only drawback of the teaching method is its limited ability to issue multiple instructions at once. The technique does not allow for the insertion of new commands to increase efficiency or adapt to changing circumstances; it simply trains the robot a predetermined set of motions.

- **Play-Back/Hand-Guiding/Lead-Through Programming**:

Lead-through programming, or hand-guiding, involves physically directing the robot arm over a sequence of points and axes to "teach" it how to accomplish a specific task. For example, a robot can be instructed by an operator on how to reach for and utilize a tool on a workstation. The robot will continue the same route on its own after it has stored the instructions in its memory.

Smaller modern robots, specifically designed to work alongside human operators, prefer hand-guiding. Indeed, the robot can train and perform intricate tasks like painting, sorting items, precision welding, and engraving, among others.

Lead-through or hand-guiding offers several advantages, which are as follows:

- Intuitiveness.
- Safety.
- Collaboration.
- Great for complex tasks or movements.

However, as a programming tool, hand guidance falls short when accuracy is needed. As such, this approach is not designed for human cooperation and is not feasible for big industrial robots. For contemporary collaborative robots utilized in smaller industrial settings, this approach is the most popular.

5.1.2 Offline-Programming Method

Using offline programming, robots or programmers may generate programs and route data straight from CAD models of the items they are handling. Writing instructions on a different system and testing them on virtual versions of industrial robots is the process of robotic programming. The written and verified instructions are uploaded into the robot's memory. Before implementing the software in real life, the manufacturer needs to create it and test a lot of times. Offline programming techniques usually work best for complicated applications where human programming would take a long time. The following are a few benefits of offline programming:

- More complexity is possible.
- More effective robot operation.
- Supports precision.
- Safety is ensured.

Due of its technical nature, offline programming is not recommended for non-engineers who are not familiar with low-level robotics programming or coding. All programming techniques are still widely used today in the majority of businesses. Depending on the type of robot and its programming, the industry can select a technique [22].

5.2 Control Techniques in Robotics

5.2.1 Force Control in Robotics

The capacity to manage the physical touch between a robot and its surroundings is a basic need for a manipulation job to be successful. Sadly, basic motion control is insufficient since modeling errors and unknowns can raise the contact force and make the interaction unstable, particularly in stiff surroundings. A robotic system needs force feedback and control in order to function reliably and safely in human presence, as well as to exhibit robust and varied behavior in unstructured situations. Maintaining a desired force or limiting the contact forces might be the control aim along the directions where interaction occurs while ensuring a desired motion in the other directions. The ability to regulate the force that a machine or a robot manipulator applies to an item or its surroundings is known as force control. During handling, regulating the contact force helps to avoid harming people, the machine, and the processed products. By preserving a constant contact force, it helps minimize wear and correct faults in manufacturing activities [23].

5.2.2 Cartesian Control

A Cartesian control robot has the ability to move a robotic manipulator arm along linear Cartesian axes. To put it simply, the robot arm can move in a straight line in the following directions: forward, back, left, right, up, and down. Certain robotic systems, referred/known as "Cartesian Coordinate Robots," are limited to moving linearly along predetermined axes. An excellent illustration of this is a CNC machine, or the claw game at your neighborhood arcade [24].

6 Software Platforms for Robot Control

Industrial robot programming and control require a symbiotic relationship between software intelligence and precise hardware. A robotics software platform is a software bundle that provides the following characteristics to simplify the programming of different robotic devices:

- an integrated programming and service execution environment;
- a collection of reusable parts;
- an environment for debugging and simulation;
- a package containing "drivers" for the majority of robotics hardware;
- a package containing common features like computer vision, navigation, or robotic arm control.

Below are the software platforms that stand out in the industry for their robustness, flexibility, and user-friendly interfaces, making them perfect for your robotic programming and control needs:

- **Robotics IDEs**: Integrated Development Environments (IDEs) are essential for writing and testing code. A good IDE for industrial robotics offers a comprehensive suite of tools tailored to the specific requirements of robot programming. It should include features like syntax highlighting, code completion, and debugging tools. Additionally, it should support the programming languages commonly used in robotics, such as Python, C + + , or proprietary scripting languages. The user interface should be intuitive, allowing users to visualize the robot's movements and operations using the provided code.
- **Simulation Software**: Before deploying code to a physical robot, simulation software can save time and reduce risk by allowing you to visualize and test the programs in a virtual environment. This type of software should accurately replicate the physics and constraints of a real-world robot. It often includes features like collision detection and real-time feedback, enabling you to refine the code and ensure it operates as intended before running it on actual hardware.
- **Motion Planning**: Motion planning software is pivotal for defining the trajectories and actions of industrial robots. The software must calculate optimal paths while avoiding obstacles and ensuring the safety of human workers. Look for platforms that allow for easy parameter adjustment and can handle complex tasks like picking, placing, and assembling with precision. The best motion planning tools offer a balance between simplicity for quick tasks and the depth needed for more intricate operations.
- **Control Frameworks**: Motion planning software is pivotal for defining the trajectories and actions of industrial robots. The software must calculate optimal paths while avoiding obstacles and ensuring the safety of human workers. Look for platforms that allow for easy parameter adjustment and can handle complex tasks like picking, placing, and assembling with precision. The best motion planning tools offer a balance between simplicity for quick tasks and the depth needed for more intricate operations.

- **Collaborative Robotics**: With the rise of collaborative robots (cobots) that work alongside humans, software platforms need to ensure safety and ease of interaction. These platforms should support features like teach pendants, which allow operators to manually guide a robot through movements and then replicate those movements automatically. The software should also include safety protocols and be compliant with industry standards to prevent accidents.
- **Customization Tools**: Lastly, customization tools are vital for tailoring the software platform as needed. Whether it's integrating with existing systems or developing custom algorithms for unique tasks, the software should offer flexibility. It should provide APIs (application programming interfaces) or SDKs (software development kits) that allow for the extension of functionality and seamless integration with other tools and systems [25].

7 Perception, Localization, and Mapping

A variety of fundamental technological components are needed by autonomous and assistive systems, such as cars, trucks, and airplanes, in order for them to operate dependably and securely. Having a strong sense of one's surroundings is crucial for many types of mobility and work environments in order to identify safe zones and prevent accidents.

Multimodal sensors are required for this, in addition to data analysis techniques like machine learning and image processing. Another approach uses visual locality determination techniques or other positioning data from satellites to determine the location on a particular map. On the other hand, map data is frequently lacking or unavailable. Research has concentrated on creating sensors and techniques for autonomous systems in a variety of application domains, location determination and mapping for support systems, and safe surroundings detection [26].

7.1 Sensor Technologies for Perception

In an artificial system, the notion of perception gives an alternative method of awareness by categorizing the pertinent aspects of an event that occurs in the surroundings [27]. The multimodal sensor data from cameras (RGB, IR, and multispectral), laser scanners, radars, time-of-flight, or ultrasonic sensors provides the basis for the perception of the environment. In order to provide techniques for data processing, analysis, and fusion that can extract pertinent information for tasks such as object categorization, tracking, localization determination and orientation, collision avoidance, and scene understanding at a higher level of abstraction, researchers performed study. In addition to 2D mapping of the environment based on point clouds produced by laser scanners, stereo camera systems, or radar sensors, 3D modeling of the surroundings is necessary for location determination, navigation, and proper item handling by machines (such as cranes or forklifts). Adverse weather conditions, such as persistent rain, haze, soil, or dust, provide a considerable obstacle. Light-based sensors cannot handle such situations. Therefore, in order to preserve crucial aspects of vision, we must substitute physical modalities such as radar. Every sensor modality has advantages and disadvantages, and combining data is required to make up for the disadvantages.

7.2 Object Recognition and Localization Methods

Accurately locating oneself and maintaining direction on an environment reference map is a prerequisite for numerous support systems as well as the dependable operation of automated machinery or vehicles. GNSS, or satellite navigation systems, commonly offer this data. Accuracy can be improved by using correction signals from reference stations supplied by the Earth. Satellite masking makes GNSS alone unreliable, especially in metropolitan areas, mountainous locations, or on corporate property. As a result, other techniques like SLAM (simultaneous localization and mapping) or local radio-based systems must be added to or replaced. For example, for vehicles and machines, precise mapping with semantic content segmentation and dynamic updating is necessary to record object-specific navigation and operational contexts. The maps can be made and altered by sensors like cameras, laser scanners, or imaging radar sensors that are mounted on equipment, cars, or infrastructural structures on the property [28].

7.3 Challenges in Robot Perception

Robots' issues in perceiving the world are not unlike from those faced by humans. First off, robots may meet scenarios where their sensors may not function as well as they should, just like humans do when they cannot see in the dark. This kind of sensory deterioration in a robotic system can be brought on by any circumstance that exceeds a sensor's operational thresholds. Whatever the reason, this leaves one without the knowledge required for accurate perception.

Challenges can develop in repetitive situations, where a lack of environmental distinguishing features can lead to confusion and make it difficult for the robot to correctly detect its location. We refer to this as perceptual aliasing when the surroundings have a maze-like appearance or resemble locations the robot has visited previously. Similar to this, when a robot receives insufficient data from its sensors to determine whether or not it is moving in relation to another item, it may experience perceptual ambiguity. According to theory, observability is the system's capacity to precisely determine its state via sensor observations.

8 Conclusion

The fundamentals of robotics are a broad field that requires both theoretical knowledge and practical experience to apply effectively. Simply put, robotics encompasses everything from robot conception to mechanical design, manufacture, and control, enabling them to perform a sequential set of tasks autonomously or semi-autonomously. Robotics is now being used in many industries, including atomic energy, automotive, defense, pharmaceutical, and textile. According to experts, there could come a time when robots surpass human intelligence. Currently, robots are able to bend an elbow, rotate the wrist, and rotate the base of an arm. It may eventually seem, feel, and behave like a human. The robot can react to its environment naturally because of its realistic-looking skin and hair.

References

1. Grimheden M, Hanson M (2005) Mechatronics—the evolution of an academic discipline in engineering education. Mechatronics 15(2):179–192
2. Wong C, Yang E, Yan XT, Gu D (2017). An overview of robotics and autonomous systems for harsh environments. In 2017 23rd International Conference on Automation and Computing (ICAC) (pp. 1–6). IEEE.
3. Bolton W (2003) Mechatronics: electronic control systems in mechanical and electrical engineering. Pearson Education
4. Ollero A, Boverie S, Goodall R, Sasiadek J, Erbe H, Zuehlke D (2006) Mechatronics, robotics and components for automation and control: IFAC milestone report. Annu Rev Control 30(1):41–54
5. Poole HH (2012) Fundamentals of robotics engineering. Springer Science & Business Media
6. Stone WL (2018) The history of robotics. In: Robotics and automation handbook (pp. 8–19). CRC Press
7. Brooks RA (1991) New approaches to robotics. Science 253(5025):1227–1232
8. https://www.seminarsonly.com/electronics/Robotics.php
9. Badaan SR, Stoianovici D (2011) Robotic systems: past, present, and future. Robotics in genitourinary surgery. Springer, London, London, pp 655–665
10. Melchiorri C, Kaneko M (2016). Robot hands. Springer Handbook of Robotics 463–480
11. Dobra A (2014) General classification of robots. Size criteria. In: 2014 23rd International Conference on Robotics in Alpe-Adria-Danube Region (RAAD) (pp. 1–6). IEEE
12. Müller-Wilderink H (2021) Automation of offline programming for assembly and welding processes in CATIA/DELMIA using VBA
13. Niku SB (2020) Introduction to robotics: analysis, control, applications. John Wiley & Sons
14. Williams B (2019) An introduction to robotics. Ohio University, Mechanical Engineering
15. McInnis BAYLISSC, Liu CK (1986) Kinematics and dynamics in robotics: a tutorial based upon classical concepts of vectorial mechanics. IEEE Journal on Robotics and Automation 2(4):181–187
16. Gasparetto A, Boscariol P, Lanzutti A, Vidoni R (2012) Trajectory planning in robotics. Math Comput Sci 6:269–279
17. Ghazaly MM, Teo TH, Vivek A, Regeev L, Kartikesu A, Vijayan L, Ali MAM (2016) Point-to-point (PTP) control performances of an upper limb robotic arm. Jurnal Teknologi 78(7)
18. Uchiyama T, Terada H, Mitsuya H (2010) Continuous path control of a 5-DOF parallel-serial hybrid robot. J Mech Sci Technol 24:47–50

19. Gupta A, Arora S, Westcott J (2016) Industrial automation and robotics: An introduction. Berlin, Boston: Mercury Learning and Information. https://doi.org/10.1515/9781683922896
20. Bär GF, Weiß G (2006) Kinematic analysis of a pentapod robot. J Geom Graph 10(2):173–182
21. Kosuge K, Furuta K (1985) Kinematic and dynamic analysis of robot arm. In: Proceedings. 1985 IEEE International Conference on Robotics and Automation vol. 2, pp. 1039–1044. IEEE
22. Zhou Z, Xiong R, Wang Y, Zhang J (2020) Advanced robot programming: A review. Current Robotics Reports 1:251–258
23. Gorinevsky DM, Formalskii AM, Shneider AY (1997) Force control of robotics systems
24. Salem FA (2015) The role of control system/algorithm subsystems in mechatronics systems design. J Multidiscip Eng Sci Technol (JMEST) 2(10): 2717–2732, *ISSN: 3159–0040, JMESTN42351005*
25. Han S, Kim MS, Park HS (2012) Open software platform for robotic services. IEEE Trans Autom Sci Eng 9(3):467–481
26. Cadena C, Carlone L, Carrillo H, Latif Y, Scaramuzza D, Neira J, Leonard JJ (2016) Past, present, and future of simultaneous localization and mapping: Toward the robust-perception age. IEEE Trans Rob 32(6):1309–1332
27. Biel L, Wide P (2000) Active perception for autonomous sensor systems. IEEE Instrum Meas Mag 3(4):27–30
28. Kefalea E (1998) Object localization and recognition for a grasping robot. In: IECON'98. Proceedings of the 24th Annual Conference of the IEEE Industrial Electronics Society (Cat. No. 98CH36200) vol. 4, pp. 2057–2062. IEEE

IOT Architecture and its Integration with Mechatronics

Rumana Tasnim, Molla Rashied Hussein, Mohammad Kamrul Hasan, Shayla Islam, Fowzia Akhter, and Ahmed Farhan

Abstract The integration of mechatronics and the Internet of Things (IoT) has appeared as one of the most vital and transformative technologies in recent years. Mechatronics, a branch of engineering, blends the principles of mechanical, electrical, electronics, and computer engineering, while IoT, a new paradigm, has profoundly altered how one interacts with technology, information, and the environment. This assimilated technology has revolutionized smart manufacturing processes in industries, utilizing mechatronic systems, automated assembly lines, wireless networks, and data analytics to optimize the entire production process in real time. Building automated systems has become faster, more effective, and more reliable than ever before by fusing the concepts of mechatronics with the seamless connectivity and real-time data processing capabilities of the IoT. The immense potential of this integrated technology has empowered engineers, researchers, and innovators worldwide to harness the power of these two. This chapter will cover the foundation of IoT systems by leveraging the key concepts for designing IoT-based mechatronic system solutions. This chapter will further explore the recent advancements and innovative breakthroughs brought by IoT, with particular emphasis on the key enabling applications for smart homes, cities, factories, healthcare, agriculture, and vehicle systems. The book delves into the potential of IOT in connecting everything in our lives and its applicability to the real world.

R. Tasnim (✉) · A. Farhan
Department of Mechatronics Engineering, World University of Bangladesh, Dhaka, Bangladesh
e-mail: rumana@mte.wub.edu.bd

M. R. Hussein
Department of Computer Science and Engineering, University of Asia Pacific, Dhaka, Bangladesh

M. K. Hasan
Center for Cyber Security, Universiti Kebangsaan Malaysia, Bangi, Malaysia

S. Islam
Institute of Computer Science and Innovation (ICSDI), UCSI University Malaysia, Kuala Lumpur, Malaysia

F. Akhter
Department of Information Technology and Engineering, Sydney Metropolitan Institute of Technology, Sydney, Australia

M. M. Rahman et al. (eds.), *Mechatronics*, Emerging Trends in Mechatronics,
https://doi.org/10.1007/978-981-97-7117-2_5

1 Introduction

The integration of IoT (Internet of Things) and mechatronics is a relatively recent development that has evolved as both fields have advanced. Mechatronics emerged in the late 1960s and early 1970s, primarily in response to the growing need for interdisciplinary approaches to engineering design. It involves the incorporation of electrical, mechanical, and computer engineering disciplines to create advanced systems with improved functionality, efficiency, and control. The concept of connecting devices and objects to the Internet to enable communication and data exchange traces back to the late twentie century. However, the term "Internet of Things" did not become widely used until the early 2000s. The evolution of wireless communication technologies, miniaturization of sensors, and advancements in data processing capabilities were crucial for the proliferation of IoT applications [1–3]. After almost 20 years of development, a growing number of academics, business leaders, and government representatives tend to think that the Internet of Things is a crucial technology for raising our standard of living and living environments. The convergence of IoT and mechatronics began to gain momentum. As IoT technology matured, it became increasingly feasible to integrate sensor networks, actuators, and data processing capabilities into mechatronic systems. This integration aimed to enhance the functionality, autonomy, and connectivity of mechatronic devices and systems. This chapter will cover the fundamentals of mechatronics and the Internet of Things, how the two systems work together, integration problems and solutions, case studies and applications, design considerations, and upcoming trends.

2 Fundamentals of Internet of Things (IoT)

We currently live in a world where continuously active, interconnected computing devices capture data capable of executing significant operations that can reshape our reality. In today's industries, characterized by numerous electronic control units and hundreds of millions of lines of code, the focus has shifted toward software rather than aerodynamics and mechanics, making the systems smarter. Smart products can perceive their surroundings, analyze large amounts of data (big data), and link to other smart devices and the Internet via a network to exchange data. Smart door locks, smart bike locks, smart kitchen appliances, irrigation controllers, smart thermostats (like Nest), Amazon Echo, and many other consumer electronics are a common sight in modern life. Nowadays, smartphones have become an integral part of our daily lives. Mobile phone prevalence is over 80% in many nations, affecting people of all ages [4]. There are more than 7 billion active phone subscribers worldwide. While Bluetooth technology frequently creates a dynamic, linked environment, it is not sufficient on its own. IoT can perform some basic actions. According to another source, there will be 24 billion IoT devices in the world by 2030, regardless of the device type [5].

The Internet of Things (IoT) refers to the process of connecting places and physical objects through smart computer devices and the Internet. It is a quickly evolving paradigm that affects many sectors of modern life, namely smart cars, homes, and more. Internet of Things (IoT) technology can make things intelligent by monitoring their surroundings, analyzing massive volumes of data (big data), and establishing Internet connection to promote data exchange. Even though smart devices have been flooding the market in recent years, there are still many prospects for researchers to create new devices and form networked systems. Consequently, the demands placed on the STEM workforce are evolving swiftly. Mechatronics engineering plays a crucial role in the innovation and design of smart devices and manufacturing systems, which are part of the Industry 4.0 revolution. As it gains popularity, engineers are becoming more and more interested in using IoT designs, processes, and resources to build IoT systems [6, 7].

Internet of Things (IoT) is a vast network of linked objects that can coordinate and share data and information on their own, as well as respond and adjust to changes in the surrounding environment. IoT communicates with the cloud and other IoT devices to share data. Gathering and sharing data from a wide range of devices paves the way for the development of automated and more effective systems. More precisely, IoT is a comprehensive and interconnected system of devices that has the capability to coordinate and exchange information, data, and assets autonomously and react and adapt to various environmental fluctuations.

IoT is an open and complete network of devices with the ability to self-organize and share information, data, and resources and respond and act in response to different types of environmental variations. We use communication technologies like radio frequency identification (RFID) and wireless sensor networks (WSN) to share data [8]. The functional elements that make up an Internet of Things system perform several functions, including sensing, actuation, communication, and management. Figure 1 illustrates the key elements of IoT below.

IoT Devices: IoT devices has the capability to engage in data exchange with other interconnected devices and applications, collect data from other devices, perform local data processing, and transmit it to centralized servers or cloud-based application back-ends for further processing. These are the actual "things" or items that have sensors, actuators, and communication modules installed in order to gather and share data with other systems or devices. Smart appliances, environmental sensors, wearable sensors, wearable fitness trackers, smart watches, cars, LED lights, industrial machinery, thermostats, etc. are a few examples of IoT devices.

IoT Gateway: IoT gateway acts as a bridge between cloud computing and IoT devices. It carries out a number of functions, including preprocessing, data filtering, protocol translation, and data aggregation. Gateways are frequently used to facilitate communication between devices that adhere to various standards or protocols.

IoT Cloud: IoT cloud processes, stores, and analyzes data gathered from IoT devices. Cloud systems offer scalable computing and storage capacity to manage massive data volumes produced by IoT devices.

Data Analytics: It is a crucial component of IoT systems responsible for extracting important insights and useful information from the vast volumes of data generated

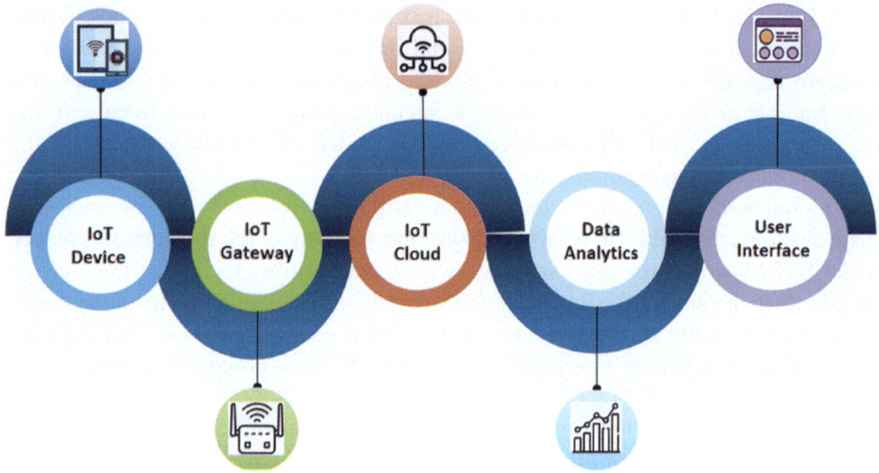

Fig. 1 Key elements of IoT

by IoT devices. In order to identify patterns, trends, and anomalies in the data, many techniques such as statistical analysis, machine learning, predictive modeling, and anomaly detection can be employed.

User Interface: Through the user interface, users may visualize data, operate connected devices, and engage with the Internet of Things. It might be in the form of desktop programs, mobile applications, or web-based dashboards. To manage their IoT devices and gain insights from data analytics, customers can access real-time monitoring, remote management, and customization options through an interface. Internet of Things (IoT) enables universal connectivity, allowing any object to establish connections with any other object, regardless of location or time, using any available network [9, 10]. Figure 2 displays the following diagram, which illustrates any connections made through IoT.

2.1 IoT Functional Block

The foundation of the Internet of Things system consists of hardware for sensing, actuating, controlling, and monitoring. Figure 3 represents the block diagram that connects the device components.

IoT Device: An Internet of Things device could have many interfaces for wired and wireless connections with other devices. Data analytics evaluate the data provided by IOT devices, producing insightful results that can guide future activities. These consist of sensor I/O interfaces, Internet connectivity interfaces, memory and storage interfaces, and audio/video interfaces. To enable their functionality, Internet of Things devices usually consist of multiple components.

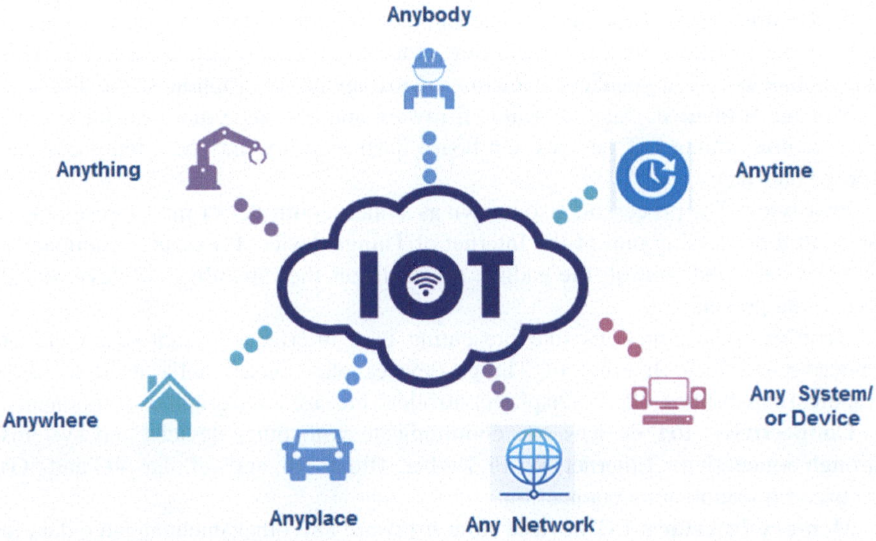

Fig. 2 Any connection through IoT

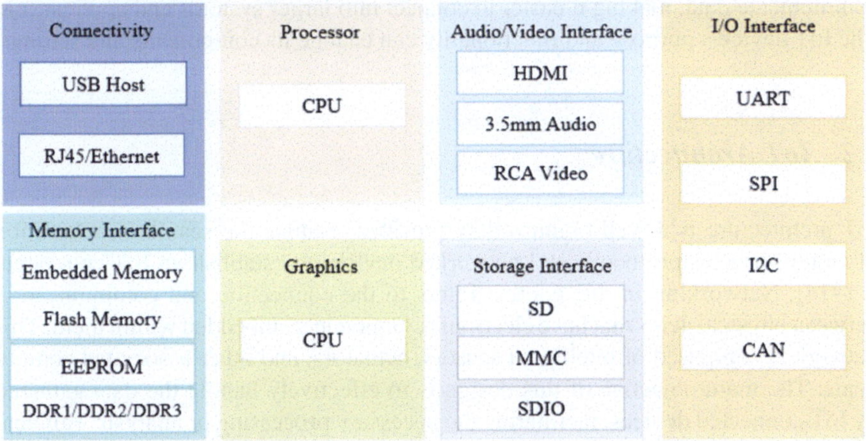

Fig. 3 Block diagram of IoT device components

I/O (Input/Output) Interface: This interface takes inputs from sensors and sends outputs to actuators, enabling the Internet of Things device to communicate with its surroundings. It consists of pins for digital and analog input and output, communication interfaces (such as SPI, I2C, CAN, and UART), and other interfaces for connecting to actuators and sensors.

Audio and Video Interface: Some Internet of Things devices may have audio and video capabilities for audio recording, video capture, or multimedia output. This may include parts like speakers, cameras, video displays, microphones, and displays.

Storage Interface: Data recording, firmware updates, and other uses for storage are common among IoT devices. On-board flash memory may be a feature of the storage interface.

Processor: The processor, also known as a microcontroller or microprocessor, is the central processing unit of the Internet of Things device. It executes commands, analyzes data, and controls the gadget. It establishes the capabilities of devices and processing power.

Graphics: For functions like presenting user interfaces, visualizing data, or rendering graphics, Internet of Things devices may occasionally need to have graphical capabilities. GPUs, Display controllers, etc. are some graphics components.

Connectivity: IoT devices can communicate with other devices and systems through connectivity. Ethernet, Wi-Fi, Zigbee, Bluetooth, and cellular (4G and 5G) are used for establishing connection.

Memory Interface: IoT devices store firmware and other unchangeable data in ROM and temporary data in RAM. Memory interfaces allow access to these types of memory.

These pieces enable the Internet of Things (IoT) gadget to sense, analyze, and communicate data, making it easier to connect into larger systems and applications. The IoT device's purpose and functionality can change its components and settings.

2.2 IoT Architecture

IoT architecture is a well-organized system that enables the seamless integration of cloud services, protocols, and networked devices to establish an IoT ecosystem [11–14]. Networking in this context refers to the connection and communication between physical items that have electronic components embedded within them. This network is composed of intelligent sensors, actuators, and other associated components. The main objective of this design is to effectively handle the data gathered by IoT-connected devices, facilitating the necessary processing or analysis. Efficient communication between sensors, devices, and network infrastructure depends on the smooth integration and interoperability of these components. After completion, the wireless sensors and smart devices of the system utilize efficient data processing techniques to facilitate specific activities such as troubleshooting any system issues. The success of Internet of Things (IoT) implementations relies on a clearly defined architecture that enables diverse components to collaborate and enhance security without introducing complications. A well-designed Internet of Things (IoT) architecture is crucial for the effective adoption in various sectors. The four separate layers, namely perception/sensing, network/connectivity, data processing, and user interface/application, are crucial. Before we can build an effective IoT system, we need to understand how each component works. In the sensing part, we look at the actual world through sensors; in the network, we can communicate analog data between different endpoints via various protocols; in data processing, we organize the raw data into meaningful insights for further analysis; and in the application level, we make all this collected detail easily actionable via a user interface or application programming interface (API). Figure 4 illustrates the layers of IoT architecture.

Perception/Sensing Layer

The sensing layer forms the foundation of an Internet of Things system. It consists of sensors and actuators that gather information from outside sources and process it. This layer plays a crucial role in collecting raw data from physical environments, enabling higher-up architectural levels to use it as input for analysis. For example, robotic functions can deploy a sensor on an automotive assembly line and relay its output to other layers for processing, enabling the assessment of quality control. Without this level in any Internet of Things network, these systems wouldn't be able to gather the essential data required to interact with reality, which would limit their potential both internally and externally.

Network/Connectivity Layer

For digital data to flow between all components of an Internet of Things system, it must pass over the network layer, sometimes referred to as the transport or device

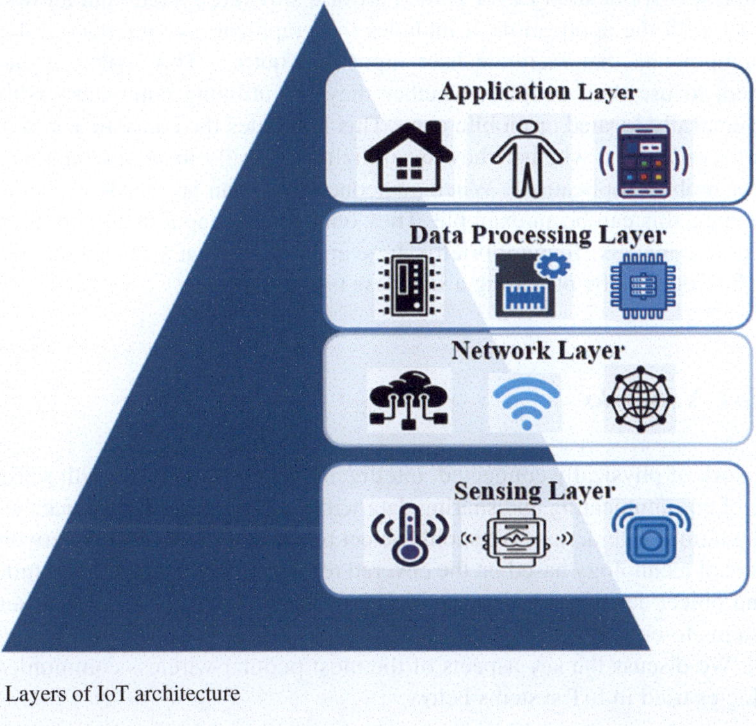

Fig. 4 Layers of IoT architecture

layer. It utilizes technologies such as HTTP, MQTT, and AMQP to facilitate communication between devices and applications. In addition, the presence of private 5G mobile networks, which provide complete control over the transmission of information, can provide a secure environment. This comprises several components: network gateways that serve as bridges to connect multiple devices concurrently; intranet ports that function as Internet gateways and access points within local systems; Internet gateways that enable communication beyond local systems; and data acquisition systems (DAS) that facilitate the gathering of sensor readings from diverse applications within a system.

Data Processing Layer

The data processing layer is accountable for examining and manipulating the gathered data to enhance decision-making and optimize operational efficiency inside the firm. The solution utilizes machine learning algorithms to analyze raw data from IoT systems, ensuring that crucial characteristics are retained for automated decision-making. This step utilizes edge analytics and artificial intelligence methods to remove irrelevant or unnecessary device information before analysis. This implementation enhances the quality of insights obtained from raw data, allowing for effective actions to be taken based on specific needs.

User Interface/Application Layer

User Interface/Application Layer is the part of a software system that allows users to interact with the application. It includes the graphical user interface (GUI) and other components that facilitate user input and output. This makes it easy for consumers to use their gadgets, whether they are utilizing computer dashboards that are centrally located or mobile apps. This facilitates the convenience of gadget usage for consumers, whether they are utilizing centrally located computer dashboards or mobile applications. When someone utilizes an app made especially for smart homes, this can be an example. They only need to tap a button on the app to turn on coffee makers. To put it briefly, these apps ensure that users get the most out of their IOT devices by providing a seamless user experience.

2.3 IoT Networks

IoT network of physically connected, intelligent objects, integrated with sensors, is capable of communicating and sharing data with other systems and devices without requiring human interaction. We should select the appropriate wireless network and IoT protocol technology based on the covered region, power usage, data volume and type, and object density. An Internet of Things network connects devices, sensors, and systems to enable data sharing, analysis, and the production of useful insights [15–18]. We discuss the key aspects of the most popular wireless communications technologies used in IoT systems below.

2.3.1 Bluetooth

Bluetooth is a wireless communication technology developed by the Bluetooth Special Interest Group (SIG) mostly used for connecting computer peripherals. Its primary function is short-distance communication between devices. The 2.4 GHz ISM band is where it operates. Bluetooth Low Energy was first introduced in 2011 and is also known as Bluetooth Smart [19]. It offers advantages over previous versions, including reduced power consumption, faster setup, and support for a number of nodes in a star topology. Low-bandwidth and self-powered devices are the target audience for Bluetooth Low Energy (BLE).

2.3.2 Near-Field Communication (NFC)

Near-field communication (NFC) is a wireless communications technology that allows bidirectional data transfer when the devices are put side by side [20–22]. NFC uses radio frequency identification and can work at frequencies up to 13.56 MHz. When the device writes data, the tag can be set to read-only or read-write; it only holds a small amount of data. Contactless payment on mobile devices, including access to all kinds of personal data, is the most prevalent use case.

2.3.3 Wi-Fi

The Institute of Electrical and Electronics Engineers (IEEE) assigned the most popular communication standard, 802.11, which is best known as Wi-Fi [23–25]. A non-profit organization known as the Wi-Fi Alliance gives certification to goods with Wi-Fi interoperability and promotes the technology. To establish a core network of this nature, both an access point and a wireless network adapter are necessary. This network runs within the frequency range of 2.4–5 GHz. One of the most important technologies currently available is Wi-Fi, which has a large user base, fast data transfer speeds, and a high market penetration rate.

2.3.4 Cellular 3G/4G/5G LTE

Cells are predetermined forms and sizes that allow radio waves to traverse a vast region; these divisions are what make up the cellular network [26, 27]. A single base station, or transceiver, equips each region, facilitating the connection of devices within or between cells. Such networks span a huge region with little power consumption. Since then, there have been several iterations, with the most notable differences being in transmission mode (analog–digital) and power consumption. Mobile phones and IoT devices, which require substantial data transfer capacities, greatly appreciate the latest cellular technologies, 5G and LTE.

2.3.5 LoRaWAN

LoRaWAN is a wireless network that is a well-known instance of low-power wide area network (LPWAN) technology. According to the LoRa Alliance, LoRaWAN provides a framework for networks and communication protocols, whereas LoRa is "just" a physical layer for wireless data transmission over great distances. LoRaWAN works with both the North American and European regulatory frameworks. In 2015, the LoRa Alliance made an announcement on the bands used for data transmission in Europe and North America. Europe operates within the 864–870 MHz ISM (Industrial, Scientific, and Medical) band, but North America operates between the 903–914.9 and 923.3–927.5 MHz ranges [28]. Many modern business and industrial solutions are based on the LoRaWAN concept.

2.3.6 NB-IoT

The LTE-IoT family of wireless networks includes the NB-IoT (narrow-band IoT) variant. This system relies on 4G mobile network functionality. Not only does NB-IoT cost less than eMTC, but it also has lower power consumption and better coverage compared to other LTE-based networks. The extensive coverage spans frequencies from 450 MHz all the way up to 3.5 GHz. Europe operates within the 864–870 MHz ISM (Industrial, Scientific, and Medical) band, but North America operates between the 903–914.9 MHz and 923.3–927.5 MHz ranges. Network providers such as Tele2, Telit, and T-Mobile, along with the founders of Ericsson, Huawei, and Nokia, established a collaborative working group with the aim of modifying and endorsing the NB-IoT technology. This group was responsible for developing the first commercial NB-IoT products [29].

2.3.7 ZigBee

The United States first introduced this network at the start of the current century. In 2003, the IEEE standardized it. The unique characteristics of the ZigBee wireless network have propelled it to the forefront of its industry. The ZigBee wireless network has stuck to 8-bit microcontrollers for all these years, even while rival networks have tried to add innovative features and boost performance. The extended gadget life is the most well-known aspect of this network. The lifespan is quite long, reaching ten years or longer in some cases. While it shares certain similarities with Bluetooth, ZigBee stands out because of its low data rate, which allows for reliable and affordable data transfer. Its properties make it ideal for sensor networks, and in high optical sight conditions, its transmission distance can reach 100 m [28].

3 Synergies Between IoT and Mechatronics

3.1 *Key Concept of Integration*

Mechatronics and the Internet of Things (IoT) integrate physical systems with digital intelligence and connection to create a potent synergy. Integration of IoT (Internet of Things) and mechatronics technologies enhances the capabilities and functionalities of both systems, resulting in substantial synergy.

Industries integrate IoT with mechatronics to develop intelligent and interconnected systems that improve performance and reliability in diverse areas such as manufacturing, transportation, healthcare, agriculture, and infrastructure. Mechatronics is the integration of mechanical systems and electronics, enhanced by the inclusion of sensors and actuators through the Internet of Things (IoT) to gather data and enable response actions. Industrial automation in mechatronics utilizes IoT sensor data to regulate machinery, monitoring variables such as temperature and pressure. IoT devices facilitate the capture and monitoring of real-time data in mechatronic systems, encompassing characteristics such as vibration and position. IoT enables engineers to remotely control and monitor equipment, facilitating troubleshooting and adjustments without the need for physical presence. This feature is particularly advantageous for maintenance tasks in industrial environments. The Internet of Things (IoT) enables seamless connection between mechatronic systems and other platforms such as supply chain management and ERP systems, resulting in increased productivity and streamlined procedures. The integration provides several benefits, such as improved operational efficiency, increased automation, real-time monitoring, predictive maintenance, and enhanced system optimization and control.

So, by fusing the capabilities of IoT and mechatronics, industries can build intelligent, networked systems with improved performance, efficiency, and reliability across a range of industries, including manufacturing, transportation, healthcare, agriculture, and smart infrastructure.

3.2 *Connectivity of Internet of Things with Mechatronics Systems*

Figure 5 illustrates the overall IoT connectivity, where the applications (Software as a Service; SaaS), platform (Platform as a Service; PAAS), and infrastructure (Infrastructure as a Service; IaaS) components all play a vital role. In addition, the system employs a blend of P2P, M2P, and M2M (person-to-person, machine-to-machine) communication [30–33]. This makes it easier for a variety of reconfigurable devices to connect to the Internet and enables the development of personalized information-based systems.

Fig. 5 IoT connectivity

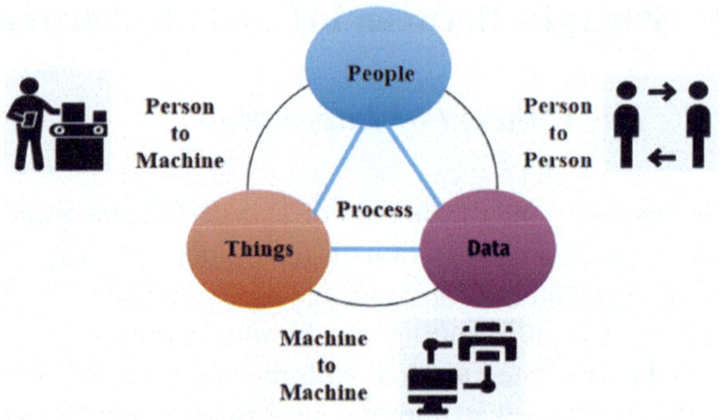

System designers have faced challenges in transitioning functionality from mechanical to electronics and information technology domains effectively, particularly in dealing with increasing complexity [34, 35]. Figure 6 illustrates how the Internet of Things (IoT) enabled intelligent objects connected to the cloud via the Internet to access and utilize cloud-stored data independently. A cloud client, in the form of a connected smart object, can access and use many services and facilities stored in the cloud [36]. Devices like tablets and smartphones, artifacts like automobiles and appliances, and components like actuators and sensors serve as smart objects that service the cloud and supply the Internet of Things' underlying connectivity. The mechatronics artifacts and smart objects in Fig. 6 are actively searching the cloud for information to help them fulfill their responsibilities. To ensure their efficacy, those objects must be positioned suitably within the environment.

The challenges faced by designers have grown because of the evolution of technology and the capacity to produce a larger amount of information. These challenges call for greater interpersonal communication as well as the integration of knowledge and understanding. Devices, as previously mentioned in the earlier sections, are also transforming the nature of systems.

Fig. 6 Cloud based system

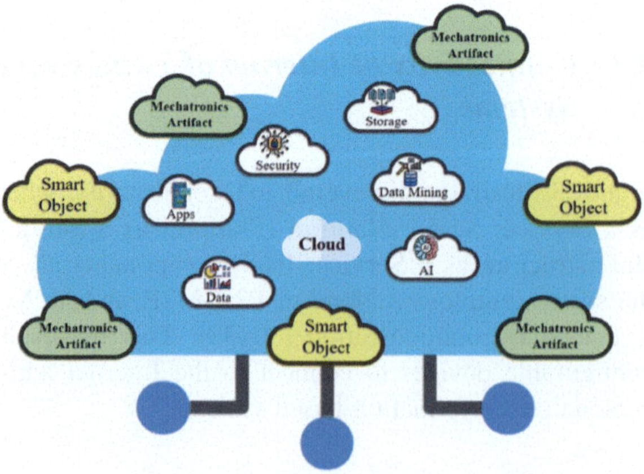

3.3 *Integration Challenges and Solutions*

The Fourth Industrial Revolution, often known as Industry 4.0, significantly depends on the Internet of Things (IoT) as a key technology that enables its advancements. The increasing popularity of user-configured and user-oriented systems, which are based on cloud-configured data and the Internet of Things, brings about new risks to the fundamental concepts of integration and transfer of functionality that form the foundation of mechatronics. Educators' and designers' responses to and management of these challenges will likely have a significant impact on the future of mechatronics and the Internet of Things [36]. The designer's task in mechatronics is to make system-level integration easier by expanding the fundamental design concepts of mechatronics, especially those related to the transfer of functionality. Once it identifies itself with the system, it can establish its position within it. The system can then adjust the device's performance by allowing device interchangeability and quick reconfiguration. According to some researchers, mechatronic components are at the base of the system composition hierarchy; they combine mechanical parts with electrical components and software to make a "smart" item that can respond to its environment. This mechatronic component operates inside an industrial automation environment that is based on the Internet of Things (IoT) framework. It does not require an additional software layer to convert its conventional interface into one that is compatible with IoT. This layer is the backbone of an Internet of Things (IoT) concept developed specifically for the industrial industry; it converts the conventional mechatronic part into an Industrial Automation Thing (IAT) [37].

4 Real-World Applications of Integrated IoT and Mechatronics Systems

The Internet of Things (IoT) streamlines, enhances, and automates processes by connecting machines, people, and data. The integration of sensors, connectivity, and AI holds great promise for improving the efficiency of various systems. Modern technological developments are largely due to the popularity of IoT solutions [38, 39]. These days, it's easy to find low-power, low-cost electronics and smart sensors. Wireless networks and newer cellular networks ensure the high bandwidth required to operate an IoT system. In recent times, advanced machine learning algorithms have the ability to rapidly examine huge amount of data. Cloud computing greatly simplifies data storage, transport, and analysis. All sectors are turning to the Internet of Things for answers, considering recent technical developments and pressing global issues. For instance, Unilever, Amazon, Microsoft, and Mercedes-Benz are just a few of the global corporations that have joined forces with "The Climate Pledge." By 2040, it aims to have achieved its lofty target of zero net carbon emissions. Businesses are putting money into energy and resource management systems that use the Internet of Things to make this happen. Moreover, the increase in the number

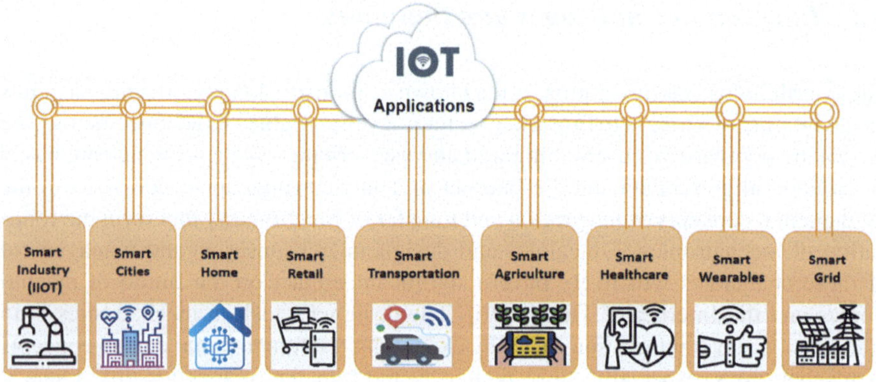

Fig. 7 IoT applications

of electric car owners has spurred investments in smart grids and smart city development. The epidemic accelerated remote work, telemedicine, Internet education, and the automation of previously laborious processes. In response to these changes, cloud service providers like AWS are investing more in the Internet of Things (IoT) cloud. These clouds provide communication services and industry-specific solutions for companies creating or utilizing Internet of Things (IoT) systems. These factors make the Internet of Things (IoT) applicable to a wide range of industries, including smart grids, smart wearables, smart agriculture, smart transportation, smart cities, smart homes, and smart industry. Figure 7 shows the real-world IoT applications.

4.1 Industrial Internet of Things (IIoT) and Mechatronics Systems for Smart Industry

The Industrial Internet of Things (IIoT) is a vast network of interconnected devices using fieldbus [40]. The created systems can immediately gather, examine, share, and act upon data to modify their surroundings or their own behavior. In real time, all of this is possible without the need for human intervention. The networking of field equipment and its potential to revolutionize manufacturing operations and efficiencies is the main emphasis of the Industrial Internet of Things (IIoT). The most crucial thing for modern manufacturing is to comprehend the intricacies of how various field devices will function within the IIoT. Fieldbus technologies facilitate communication between various field devices. The devices encompass a range of components such as sensors, actuators equipped with embedded technology, programmable logic controllers (PLCs), control stations, and supervisory control and data acquisition/human–machine interface (SCADA/HMI) devices. Additionally, industrial connectivity is now outside of the facility. Examples of IIoT applications that extend beyond the plant include wind farms, smart grids, and the monitoring of vital truck components in mines. Figure 8 depicts the industrial setting as the key focus of the IIoT.

Let's consider the Internet of Everything as a network of networks, with billions or even trillions of connections between objects. These links bring with them new security and other concerns, in addition to previously unheard-of opportunities. The integration of IIoT and industrial robots is significantly changing manufacturing and industrial processes. Furthermore, IoT digital twins are a potent example of how mechatronics technologies are coming together to allow industries to better understand their operations, maximize productivity, and spur innovation in industrial settings. "Digital twins" are virtual replicas of real-world systems or objects. In the context of the IIoT, people frequently construct a digital twin of a physical asset or process to enable real-time monitoring, analysis, and simulation. We can use this virtual model to simulate various scenarios, forecast maintenance requirements, and maximize performance. Manufacturing, energy, transportation, and healthcare are just a few of the industries that stand to gain greatly from the integration of IoT and digital twins [41]. Through the utilization of sensor data dispersed across the industrial setting, entities can fabricate precise digital twins that provide insight into machinery functionality, workflow effectiveness, and overall operations.

IIoT integrated systems provide real-time updates on the mechanical state of industrial machinery. Users also use it to remotely alter settings, activate gear, or halt operations as needed. This feature is suitable in cases where it is difficult, costly, and time-consuming to send a specialist to a remote location, especially when all that is required is a basic reboot or configuration adjustment. Another crucial use of IIoT is its capacity to track the quality of manufactured goods at every stage. Also, IIoT technologies make it possible for maintenance personnel to remotely monitor and diagnose equipment, resolving issues and identifying problems without the need

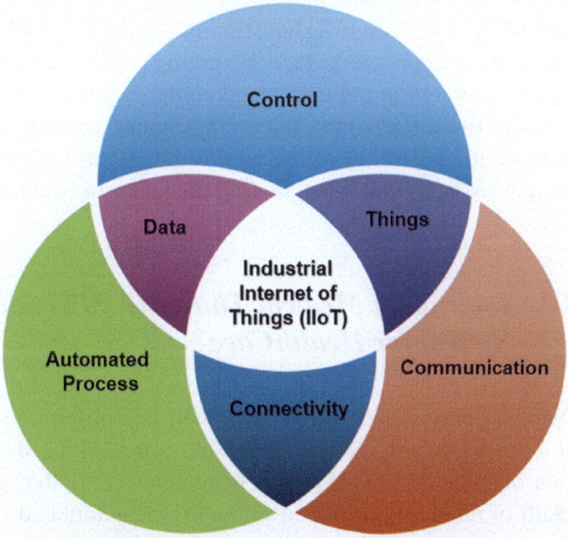

Fig. 8 Industrial Internet of Things (IIoT)

for physical presence at the location. When a machine requires maintenance, we immediately stop production before a crisis arises. Thus, it serves as one of the justifications for putting in place a system for the collection, analysis, and administration of data.

Using sensors to deliver alerts when specific risk factors appear after the installation of equipment and operating platforms is one of the most effective industrial IoT applications. For example, the oil and gas industries use IIoT with smart sensors. Sensors immediately alert the maintenance teams when they detect oil or gas leakage in any of their pipelines. This helps avert any dangers and ensures a steady supply at all times. Mechatronics systems and automated industries have recently applied industrial IoT and machine learning to precisely provide feedback on required maintenance, thereby increasing overall production and efficiency [42]. For instance, the platforms receive data from the sensors on robots or other industrial machinery. The sensors analyze the data in real time and use sophisticated algorithms to issue warnings about abnormally high temperatures or vibrations. Industrial IoT systems enable automated inventory monitoring, verifying plan compliance, and issuing alerts when it fails. Maintaining a consistent and effective workflow requires the use of this additional industrial IoT application. Using automated warnings and event reporting, industries can adjust to disturbances in the supply chain and inventories. Downtime is minimized by keeping necessary consumables on hand due to the real-time insights provided by IoT sensors. Additionally, the ability to track the state of individual components enables businesses and their clients to arrange replacement orders in advance. Furthermore, IoT-enabled machinery can identify and handle safety events by integrating IIoT with emergency response systems. For example, IIoT sensors can detect gas leaks or fires, set off alerts, and start emergency shutdown procedures to stop the situation from getting worse. Also, smart wearables integrated with IoT capabilities can track temperature, humidity, and dangerous gas concentrations, among other environmental factors. They may also monitor vital signs and identify indicators of tension or exhaustion in employees, which helps to avert mishaps and guarantee a secure workplace. By giving workers and supervisors access to real-time data on production indicators and equipment status, wearables can help them make well-informed decisions. Integrating smart sensors into tools and machinery can enhance the efficiency of the workforce. Smart glasses and wearables with specific designs assist workers in reducing mistakes and enhancing workplace security.

4.2 Internet of Medical Things (IoMT) and Mechatronics System for HealthCare

Over the past decade, there has been a proliferating interest in the development of a precise health monitoring system for tracking the movements and other activities of patients [43–45]. The healthcare sector has completely transformed as a result of emerging trends in miniaturized automated mechatronics devices and the

Internet of Medical Things (IoMT) or Medical Internet of Things (MIoT). Integration of the Internet of Medical Things (IoMT) with mechatronics systems is becoming more popular in the healthcare industry, transforming medical operations and patient care. Healthcare providers can collect real-time health data and monitor patients remotely by combining the capabilities of mechatronics and IoMT. Additionally, this connectivity improves the efficiency of healthcare procedures and lessens the burden on healthcare professionals. Furthermore, this integrated technology may lead to early identification of possible medical device failures, ensuring accurate readings and lowering equipment costs. A significant application is wearable Internet of Things (IoT) gadgets that allow hospitals to track patients' vitals while they're at home, which shortens patients' hospital stays while simultaneously giving life-saving information in real time. Smart beds in hospitals notify personnel of available spaces, reducing the amount of time patients have to wait for a bed. Sensors can detect if a patient is having a heart attack or has fallen [46–48]. The installation of Internet of Things (IoT) sensors reduces the likelihood of failure and increases dependability, making caring for the elderly much easier. With the integrated capabilities of mechatronics and IoMT, implantable drug delivery systems have advanced to a new level. Drug delivery systems may now distribute drugs with improved accuracy and control, as well as safe and focused pharmaceutical administration [49]. Implantable medical devices with mechatronic components and Internet of Medical Things (IoMT) connectivity, such as neurostimulators and pacemakers, are now available. Robotic exoskeletons and rehabilitation equipment incorporate mechatronics to assist patients with movement problems [50]. IoMT connectivity allows for remote rehabilitation sessions. Telemedicine platforms integrate mechatronics and IoMT to facilitate secure data transmission, remote patient health metrics monitoring, and high-quality video communication.

4.3 Mechatronics Systems and IoT in Smart Cities and Homes

The idea of smart cities has drawn a lot of attention in today's age of rapid technological advancement [51–53]. These cities build automated mechatronics systems that improve the effectiveness, sustainability, and livability of urban areas by utilizing a variety of technologies, such as mechatronics and the Internet of Things. Smart cities can benefit significantly from increased automation, data-driven decision-making, and connection by integrating IoT and mechatronics in smart city applications. By embedding IoT in automated mechatronic systems, which can oversee inspection procedures and even perform small repairs, we can achieve real-time condition monitoring of infrastructure elements such as buildings, roads, and bridges. Traffic lights, cameras, and sensors with IoT capabilities can improve safety, ease traffic, and minimize congestion. Using image processing and artificial intelligence techniques, mechatronic systems could control autonomous cars and adaptive traffic lights to improve traffic flow [54]. IoT-integrated smart meters can track energy consumption in buildings, lamps, and other civic infrastructure. Mechatronic systems also use IoT-integrated pumps and valves to control water flow, resulting in more effective distribution and a decrease in water waste [55]. Wi-Fi connectivity enables remote control of smart home appliances such as lighting, door locks, cameras, and thermostats using smartphones or voice commands. IoT-enabled smart bins can minimize overflow, optimize garbage collection routes, and even condense trash when necessary. Smart home systems can automate various functions based on predetermined criteria. For instance, anyone may program their thermostat to change the temperature while they are away from home or to turn on the lights automatically when they sense motion. By optimizing energy usage, IoT-enabled gadgets assist homes in saving energy and lowering utility bills.

4.4 Automated Mechatronics Systems and IoT in Smart Retail

The emergence of cyber-physical systems and the Fourth Industrial Revolution have made mechatronics ubiquitous in "smart industry." By laying the groundwork for automated manufacturing facilities and the Factory of Things, the convergence of cloud computing and the Internet of Things (IoT) has revolutionized the smart industry. Smart retail market components include smartphones, 4G and 5G networks, public Wi-Fi, and cloud computing in addition to IoT. In the conventional retail system, the manual order placement method poses a challenge for the vendor to earn the trust of the customer. However, in e-commerce, the effectiveness of the logistics has a sole bearing on the increase in sales. The smart grocery method, which uses a sensor-based marketing paradigm, places sensor-based measurement containers (SBMCs) at each subscriber site [56]. IoT-enabled sensors and actuators send data to the smart grocery system on their own, helping a partially automated grocery store business process work and figure out when to reorder a certain amount of goods.

4.5 Automated Mechatronics Systems and IoT Integration in Intelligent Transportation

An important factor propelling the worldwide IoT is the Intelligent Transportation System (ITS) [57–59]. IoT-based vehicle-to-vehicle communication has ushered in a new era of communication that will lead to ITS. In order to achieve and support efficient traffic system management, the Internet of Things combines sensor data storage, processing, and analysis with data analytics computation. By automating roads, trains, planes, and ships, IoT-based intelligent transportation systems (IoT-ITS) improve consumer satisfaction with smooth tracking, delivery, and transfer of commodities.

4.6 Automated Mechatronics Systems and IoT in Smart Agriculture and Aquaculture

The integration of IoT and sophisticated mechatronic devices into the field has significantly increased agriculture production [60, 61]. Smart mechatronics systems are becoming increasingly important in the agriculture sector because of their efficiency-related benefits and other advantages. The integration of cutting-edge technologies such as robotics, artificial intelligence, and the Internet of Things (IoT) into agricultural operations can enable robot- and drone-based farming. Farmers are able to monitor and control numerous parts of their farming operations remotely and in real time by combining the precision and efficiency of mechatronics systems with the connectivity and data processing capabilities of the IoT. This integration allows for smart irrigation, automated seeding and weeding, automated pesticide application, and real-time monitoring. The precise and strategic application of resources such as water, fertilizer, and pesticides enables cost reduction and environmental sustainability. Moreover, the application of IoT sensors and devices in smart agriculture enables constant monitoring of environmental parameters such as soil moisture, temperature, humidity, and crop growth rates. Making educated judgments about nutrient application, pest control, and irrigation schedules is possible with the analysis of this real-time data. Finally, the farming industry has undergone a revolution in terms of process optimization and productivity gains from the integration of IoT and automated mechatronics systems in smart agriculture.

The integration of mechatronics systems with the Internet of Things (IoT) has transformed fish farming in aquaculture [62–65]. Due to advancements in mechatronics engineering, aquaculture operations now use sophisticated and automated systems. Sensors, actuators, and control systems monitor and precisely regulate water quality, pH, dissolved oxygen, and temperature in real time, guaranteeing aquatic life thrives. The seamless integration of IoT technologies, which collect, analyze, and distribute data, has transformed aquaculture farmers' decision-making and operational optimization. One of the biggest advances in this discipline is IoT-integrated intelligent feeding systems [66–68].

4.7 Smart Automated Systems and IoT in Smart Grid

Energy management has been a serious challenge in recent decades due to declining energy reserves. Researchers have been working on developing real-time energy monitoring and control systems [69]. Technological advancements have transformed energy management, with mechatronics and the IoT enabling smart grids. Intelligent systems with IoT are able to monitor, analyze, and optimize grid energy consumption. One application is smart meters [70], which monitor energy consumption at specific times and report back. Meters use this information to optimize energy consumption during periods of low demand. It also manages the various energy demand spikes during specific times. Smart energy systems use newer, more precise, dynamic range, and dependable sensing technology across several domains. Communication methods have become fast, adaptable, and dynamic. Integration, powerful processing engines, and flexible, customizable software stacks have revolutionized computation. System integration has advanced through automation and data analytics paradigms. As a result, smart energy systems have been revolutionized enormously. The measurement component of smart meters for smart grids has scaled up to meet the needs of the evolving energy ecosystem using the Internet of Things (IoT) and smart meters [71–73]. By using smart meters and IoT devices in homes and businesses, utilities can optimize energy consumption during peak hours, reducing strain on the smart grid and infrastructure replacement costs.

5 Conclusion

Since its inception, the mechatronics system has undergone transformations as a result of technological breakthroughs and evolving system design ideas. However, this has led to some variation in the perception of mechatronics, particularly in the context of mechatronics education. The advent of the Internet of Things has prompted educators, practitioners, and mechatronics designers to reassess the conceptualization, development, and construction of mechatronic systems and components. The increasing focus is on topics such as machine ethics, human engagement, complexity, and context, as well as data and personal security. This is due to the role of mechatronic smart objects in IoT-based systems, where the structure is defined by context. The Internet of Things is significantly facilitating the collection of vast quantities of data from multiple sources. However, the work of collecting, managing, and interpreting the information becomes more complex due to the large volume of data generated by an unlimited number of IoT devices. The fields of artificial intelligence (AI) and the Internet of Things (IoT) provide significant potential for advancement and concrete benefits in this regard.

References

1. Gubbi J, Buyya R, Marusic S, Palaniswami M (2013) Internet of Things (IoT): a vision, architectural elements, and future directions. Futur Gener Comput Syst 29(7):1645–1660

2. Wang J, Lim MK, Wang C, Tseng ML (2021) The evolution of the Internet of Things (IoT) over the past 20 years. Comput Ind Eng 155:107174
3. Harald Sundmaeker PF, Guillemin P, Woelfflé S (2010) Vision and challenges for realising the Internet of Things. Pub Office EU
4. Statista (2022) Number of smartphone users worldwide. www.statista.com/statistics/330695/number-of-smartphone-users-worldwide
5. Transforma Insights (2020) Global IoT market to grow to 24.1 billion devices in 2030, generating $1.5 trillion annual revenue
6. Ahmed SA, Alwan NF, Ali AM (2018) Overview for Internet of Things: basics, components and applications. J Univ Anbar Pure Sci 12(3)
7. Madakam S, Ramaswamy R, Tripathi S (2015) Internet of Things (IoT): a literature review. J Comput Commun 3:164–173
8. Khan R, Khan SU, Zaheer R, Khan S (2012) Future internet: the internet of things architecture, possible applications and key challenges. In: Proceeding of 10th international conference on frontiers of information technology, Islamabad, pp 257–260
9. Ghasempour A (2019) Internet of things in smart grid: architecture, applications, services, key technologies, and challenges. Inventions 4(1):22
10. Krishnamoorthy N, Kalaimagal R, Shankar SG, Asif NA (2018) IoT based smart door locks. Int J Future Revolution Comput Sci Commun Eng 4(3):151–154
11. Meneghello F, Calore M, Zucchetto D, Polese M, Zanella A (2019) IoT: Internet of threats? a survey of practical security vulnerabilities in real IoT devices. IEEE Internet Things J 6(5):8182–8201
12. Weyrich M, Ebert C (2016) Reference architectures for the internet of things. IEEE Softw 33(1):112–116
13. Sebastian S, Ray PP (2015) Development of IoT invasive architecture for complying with health of home. In: Proceedings of I3CS, Shillong, pp 79–83
14. Aiello M (2022) IoT architectures: from data to smart systems. Front Internet Things 1:959268
15. Ray PP (2018) A survey on Internet of Things architectures. J King Saud Univ Comput Inf Sci 30(3):291–319
16. Sinha RS, Wei Y, Hwang S-H (2017) A survey on LPWA technology: LoRa and NB-IoT. In: ICT The Korean Institute of communications information sciences. Publishing Services by Elsevier B.V., pp 14–21
17. Mekkia K, Bajica E, Chaxela F, Meyerb F (2019) A comparative study of LPWAN technologies for large-scale IoT deployment. In: The Korean Institute of communications and information sciences (KICS). Publishing Services by Elsevier B.V., pp 1–7
18. Mohamed KS (2019) The era of Internet of Things towards a smart world. Springer Nature Switzerland AG
19. Sikimić M, Amović M, Vujović V, Suknović B, Manjak D (2020) An overview of wireless technologies for IoT network. In: 2020 19th international symposium INFOTEH-JAHORINA (INFOTEH). IEEE, pp 1–6
20. Gomez C, Oller J, Paradells J (2012) Overview and evaluation of bluetooth low energy: an emerging low-power wireless technology. Sensors 12(12):11734–11753
21. Motlagh NH (2012) Near field communication (NFC) a technical overview. University of VAASA, Faculty of Technology, Telecommunication Engineering
22. Bilginer B, Ljunggren P (2020) An introduction to near field communication. https://lup.lub.lu.se/luur/download?func=downloadFile&recordOId=2274992&fileOId=3629824
23. Curran K, Millar A, Garvey CM (2012) Near field communication. Int J Electr Comput Eng (IJECE) 2(3):371–382
24. Wireless Technologies for IoT (2020) https://iot.electronicsforu.com/expert-opinion/wireless-technologies-iot/
25. Song S, Issac B (2014) Analysis of Wifi and Wimax and wireless network coexistence. Int J Comput Netw Commun (IJCNC) 6(6)
26. Choose the Right IoT Wireless Connectivity Protocol. https://abracon.com/news/choose-the-right-iot-wireless-connectivityprotocol

27. Obinnaa O, Palaniappana S, Abubakara A (2014) Overview of cellular network: impact of GSM to the economic growth in Nigeria. J Adv Appl Sci (JAAS) 2(6):192–204
28. Singh RK, Singh R (2016) 4G LTE cellular technology: network architecture and mobile standards. Int J Emerg Res Manage Technol 5(12)
29. Shah SH, Yaqoob I (2016) A survey: Internet of Things (IOT) technologies, applications and challenges. In: 4th IEEE international conference on smart energy grid engineering, pp 381–385
30. Mazhelis O, Tyrväinen P (2014) A framework for evaluating internet-of-things platforms: application provider viewpoint. In: Proceedings of IEEE world forum on internet of things, pp 147–152
31. Bandyopadhyay D, Sen J (2011) Internet of things: applications and challenges in technology and standardization. Wirel Pers Commun 58:49–69
32. Moreno-Vozmediano R, Montero R, Llorente IM (2013) Key challenges in cloud computing enabling the future internet of services. IEEE Internet Comput 18–25
33. Carlin S, Curran K (2012) Cloud computing technologies. Intl J Cloud Comput Serv Sci 1(2):59–65
34. Vasić VS, Lazarević MP (2008) Standard industrial guideline for mechatronic product design. FME Trans 36:103–108
35. Bradley DA, Russell D (eds) (2010) Mechatronics in action. Springer
36. Bradley D, Russell D, Ferguson I, Isaacs J, MacLeod A, White R (2015) The Internet of Things—the future or the end of mechatronics. Mechatronics 27:57–74
37. Foradis T, Thramboulidis K (2016) From mechatronic components to industrial automation things. Web
38. Chen S, Xu H, Liu D, Hu B, Wang H (2014) A vision of IoT: applications, challenges, and opportunities with china perspective. IEEE Internet Things J 1(4):349–359
39. Udoh IS, Kotonya G (2018) Developing IoT applications: challenges and frameworks. IET Cyber-Phys Syst Theory Appl 3(2):65–72
40. Stankovski S, Ostojić G, Zhang X (2016) Influence of industrial internet of things on mechatronics. J Mechatron Autom Identif Technol 1(1):1–6
41. Kamath V, Morgan J, Ali MI (2020) Industrial IoT and digital twins for a smart factory: an open source toolkit for application design and benchmarking. In: 2020 global internet of things summit (GIoTS). IEEE, pp 1–6
42. Elkateb S, Métwalli A, Shendy A, Abu-Elanien AE (2024) Machine learning and IoT–based predictive maintenance approach for industrial applications. Alex Eng J 88:298–309
43. Arshad A, Khan S, Alam AZ, Ahmad FI, Tasnim R (2014) A study on health monitoring system: recent advancements. IIUM Eng J 15(2)
44. Teixeira E, Fonseca H, Diniz-Sousa F, Veras L, Boppre G, Oliveira J, Marques-Aleixo I (2021) Wearable devices for physical activity and healthcare monitoring in elderly people: a critical review. Geriatrics 6(2):38
45. Arshad A, Alam AHMZ, Khan S, Tasnim R (2015) Passive monitoring techniques for elderly: a review on recent development and prospective. In: Proceedings of the world congress on engineering and computer science (WCECS 2015), vol 1
46. Arshad A, Khan S, Alam AZ, Tasnim R, Boby RI (2016) Health and wellness monitoring of elderly people using intelligent sensing technique. In: 2016 international conference on computer and communication engineering (ICCCE). IEEE, pp 231–235
47. Al-Khafajiy M, Baker T, Chalmers C, Asim M, Kolivand H, Fahim M, Waraich A (2019) Remote health monitoring of elderly through wearable sensors. Multimed Tools Appl 78(17):24681–24706
48. Arshad A, Khan S, Alam AZ, Tasnim R, Gunawan TS, Ahmad R, Nataraj C (2016) An activity monitoring system for senior citizens living independently using capacitive sensing technique. In: 2016 IEEE international instrumentation and measurement technology conference proceedings. IEEE, pp 1–6
49. Cianchetti M, Laschi C, Menciassi A, Dario P (2018) Biomedical applications of soft robotics. Nat Rev Mater 3(6):143–153

50. Atashzar SF, Carriere J, Tavakoli M (2021) How can intelligent robots and smart mechatronic modules facilitate remote assessment, assistance, and rehabilitation for isolated adults with neuro-musculoskeletal conditions? Front Robot AI 8:610529
51. Kanagachidambaresan GR (2021) IoT projects in smart city infrastructure. Role of Single Board Computers (SBCs) in rapid IoT prototyping. Springer International Publishing, Cham, pp 199–215
52. Whaiduzzaman M, Barros A, Chanda M, Barman S, Sultana T, Rahman MS, Fidge C (2022) A review of emerging technologies for IoT-based smart cities. Sensors 22(23):9271
53. Syed AS, Sierra-Sosa D, Kumar A, Elmaghraby A (2021) IoT in smart cities: a survey of technologies, practices and challenges. Smart Cities 4(2):429–475
54. Shaikat AS, Saleheen RU, Tasnim R, Mahmud R, Mahbub F, Islam T (2019) An image processing and artificial intelligence based traffic signal control system of Dhaka. In: 2019 Asia pacific conference on research in industrial and systems engineering (APCoRISE). IEEE, pp 1–6
55. Tasnim R, Hussein MR, Farhan A, Saleheen RS, Zonayed M, Huq E, Mahbub F, Rahman MM (2023) IoT and GSM integrated automated water pump controlling system for prevention of water wastage. In: Proceedings of the 6th industrial engineering and operations management Bangladesh conference Dhaka, Bangladesh
56. Kaur J, Kaur PD (2018) CE-GMS: a cloud IoT-enabled grocery management system. Electron Commer Res Appl 28:63–72
57. Básaca-Preciado LC, Orozco-Garcia NA, Terrazas-Gaynor JM, Moreno-Partida AS, Rosete-Beas OA, Rizzo-Aguirre J, Ponce-Camacho MA (2018) Intelligent transportation scheme for autonomous vehicle in smart campus. In: IECON 2018–44th annual conference of the IEEE industrial electronics society. IEEE, pp 3193–3199
58. Muthuramalingam S, Bharathi A, Rakesh Kumar S, Gayathri N, Sathiyaraj R, Balamurugan B (2019) IoT based intelligent transportation system (IoT-ITS) for global perspective: a case study. Internet Things Big Data Anal Smart Gener 279–300
59. Aldakkhelallah A, Simic M (2021) Autonomous vehicles in intelligent transportation systems. In: Human centred intelligent systems: proceedings of KES-HCIS 2021 conference. Springer Singapore, pp 185–198
60. García L, Parra L, Jimenez JM, Lloret J, Lorenz P (2020) IoT-based smart irrigation systems: an overview on the recent trends on sensors and IoT systems for irrigation in precision agriculture. Sensors 20(4):1042
61. Amertet S, Gebresenbet G, Alwan HM, Vladmirovna KO (2023) Assessment of smart mechatronics applications in agriculture: a review. Appl Sci 13(12):7315
62. Chiu MC, Yan WM, Bhat SA, Huang NF (2022) Development of smart aquaculture farm management system using IoT and AI-based surrogate models. J Agric Food Res 9, 100357
63. Balakrishnan S, Rani SS, Ramya KC (2019) Design and development of IoT based smart aquaculture system in a cloud environment. Int J Oceans Oceanogr 13(1):121–127
64. Tsai KL, Chen LW, Yang LJ, Shiu HJ, Chen HW (2022) IoT based smart aquaculture system with automatic aerating and water quality monitoring. J Internet Technol 23(1):177–184
65. Tasnim R, Shaikat AS, Al Amin A, Hussein MR, Rahman MM (2022) Design of a smart biofloc monitoring and controlling system using IoT. J Eng Adv 3(04):155–161
66. Abu-Khadrah A, Issa GF, Aslam S, Shahzad M, Ateeq K, Hussain M (2022) IoT based smart fish-feeder and monitoring system. In: 2022 international conference on business analytics for technology and security (ICBATS). IEEE, pp 1–4
67. Auliar R, Bekaroo G (2020) A smart fish feeding system for internet of things based aquariums. In: Proceedings of the 2nd international conference on intelligent and innovative computing applications, pp 1–8
68. Al-Mutairi AW, Al-Aubidy KM (2023) IoT-based smart monitoring and management system for fish farming. Bull Electr Eng Inf 12(3):1435–1446
69. Shaikat AS, Tasnim R, Saleheen RU, Rayhan M, Khan MT, Hasan R (2019) A real time electrical load distribution monitoring and controlling system based on PLC and webserver. In: 2019 international conference on energy and power engineering (ICEPE). IEEE, pp 1–6

70. Morello R, De Capua C, Fulco G, Mukhopadhyay SC (2017) A smart power meter to monitor energy flow in smart grids: the role of advanced sensing and IoT in the electric grid of the future. IEEE Sens J 17(23):7828–7837
71. Rind YM, Raza MH, Zubair M, Mehmood MQ, Massoud Y (2023) Smart energy meters for smart grids, an internet of things perspective. Energies 16(4):1974
72. Siozios K, Anagnostos D, Soudris D, Kosmatopoulos E (2019) IoT for smart grids. Springer, Cham, Switzerland, p 30
73. Bordin C, Håkansson A, Mishra S (2020) Smart energy and power systems modelling: an IoT and cyber-physical systems perspective, in the context of energy informatics. Procedia Comput Sci 176:2254–2263

Machine Learning and Artificial Intelligence

Md. Sadatuzzaman Saagoto, Ahmed Farhan, Protik Barua, Ifthekhar Hussain, Kamrul Hasan, and Tamanna Hasan

Abstract Artificial intelligence is a very powerful tool for engineers. Mechatronics can leverage AI and ML to bring revolutionary changes in different domains of this vast subject. Control engineering, robotics and automation, autonomous vehicles, medical devices, quality control, precision agriculture, and remote sensing can all benefit from the implementation of AI and ML. The first part of this chapter will focus on the fundamentals of machine learning, artificial intelligence, and various techniques used in ML, AI, and deep learning. The second part will deal with AI and ML in control systems, robotics, and automation. This chapter will briefly discuss the challenges of implementing AI and ML in mechatronics systems. This chapter focuses on the development of a knowledge of ML and AI approaches utilized in mechatronics engineering.

Keywords Machine learning · Artificial intelligence · Robotics · Supervised learning · Unsupervised learning

1 Introduction

Artificial intelligence (AI) is the aptitude of a computer or robot controlled by a computer capable of performing activities that are frequently linked with the people of intelligence. The phrase is widely used to refer to the attempt of assembling systems that have human-like intellectual processes, such as the reasoning ability, distinguish meaning, or learn from previous encounters. On the other hand, machine learning

Md. Sadatuzzaman Saagoto (✉) · A. Farhan · P. Barua · I. Hussain · K. Hasan
Department of Mechatronics Engineering, World University of Bangladesh, Dhaka, Bangladesh
e-mail: saagoto57@mte.wub.edu.bd

A. Farhan · P. Barua · I. Hussain · K. Hasan
Department of Mechatronics and Industrial Engineering, Chittagong University of Engineering and Technology, Chittagong, Bangladesh

T. Hasan
Department of Textile Engineering, World University of Bangladesh, Dhaka, Bangladesh

M. M. Rahman et al. (eds.), *Mechatronics*, Emerging Trends in Mechatronics,
https://doi.org/10.1007/978-981-97-7117-2_6

(ML) is a discipline of AI that concentrates on the development and investigation of statistical algorithms that can derive information from data and make inferences from previously unknown data, allowing them to complete tasks without specific guidance. In allowing computers to do tasks that require human supervision traditionally, the fast-developing sciences of artificial intelligence (AI) and machine learning (ML) are changing a variety of industries. AI covers a wide spectrum of technologies, from limited AI, which is best at certain jobs, to the speculative idea of AI, which would be better than humans. Algorithms used in ML, a branch of AI, use supervised, unsupervised, and reinforcement learning approaches to study from data and acquire more accuracy over time. Prediction analytics, driverless cars, tailored corrections, image and voice recognition, and more are uses of ML and AI. But new technologies additionally create significant social and ethical challenges, such as bias, privacy, loss of jobs, and security; therefore, it will take constant work to guarantee the responsible development and application of these advances.

2 Fundamental Concept of Machine Learning

Machine learning is an automated data analysis technique. It is an area of computer science that deals with the automatic finding of patterns in data and, based on these patterns, predicting the future or performing decision-making by applying statistical methods. Machine learning allows a system, without explicitly programming it, to learn from data or examples [1].

2.1 Workflow or Operation of Machine Learning

Before we go for the specifics of Machine Learning (ML), let's have a look at its fundamental methodology. Figure 1 illustrates the workflow of ML process that consists of several integral parts. They are exploration and data processing, training and testing data, modeling, and evaluation.

The model of machine learning is a piece of smart code that is formed by testing and training gathered data. Data gathering is subjected to on the category of projects. Different types of sensors, databases, files, and Internet sources can be used as the main sources of data. After that, the raw data should be cleaned through the data processing method. Then, the analyzers have to design the best possible model to test and train the data appropriately. Finally, the model must be evaluated which is a vital part of the development process of the designed model.

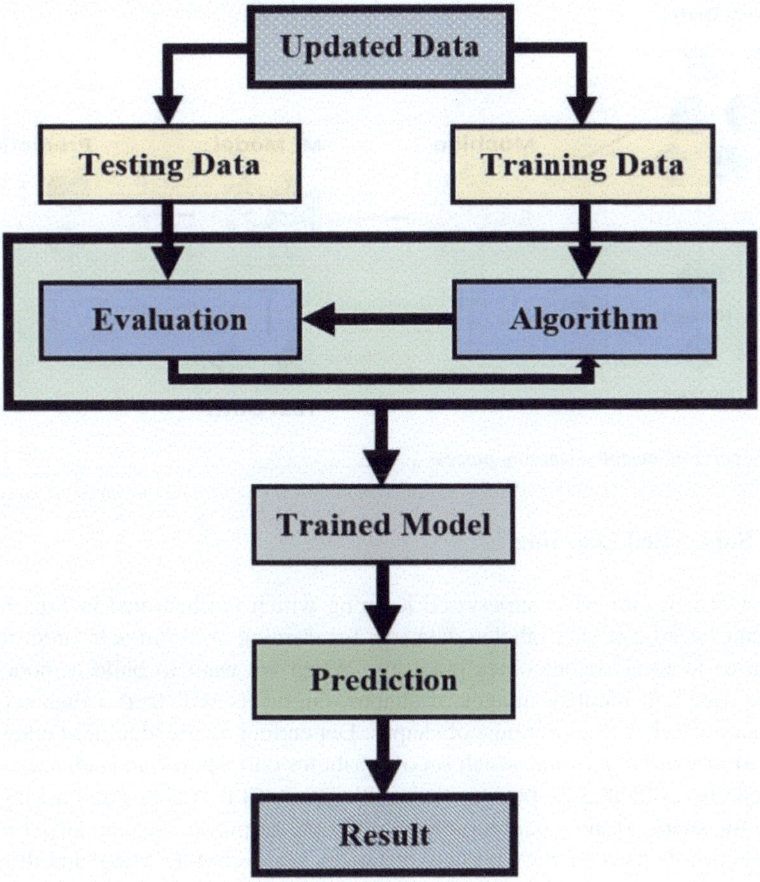

Fig. 1 Workflow of machine learning

2.2 Types of Machine Learning

Machine Learning (ML) is broadly classified into four groups: supervised, semi-supervised, unsupervised, and reinforcement learning [2]. Figure 2 shows the categories of this process.

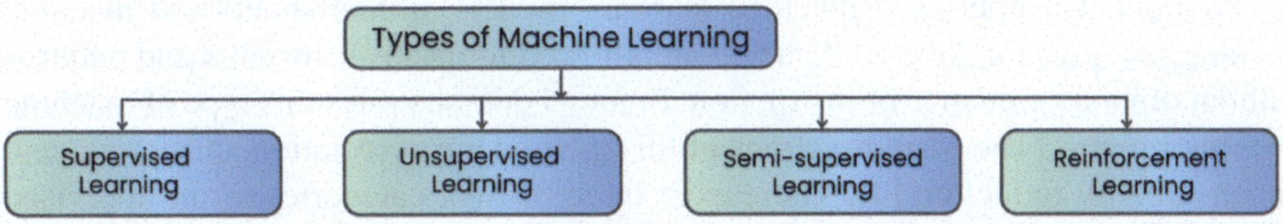

Fig. 2 Types of machine learning

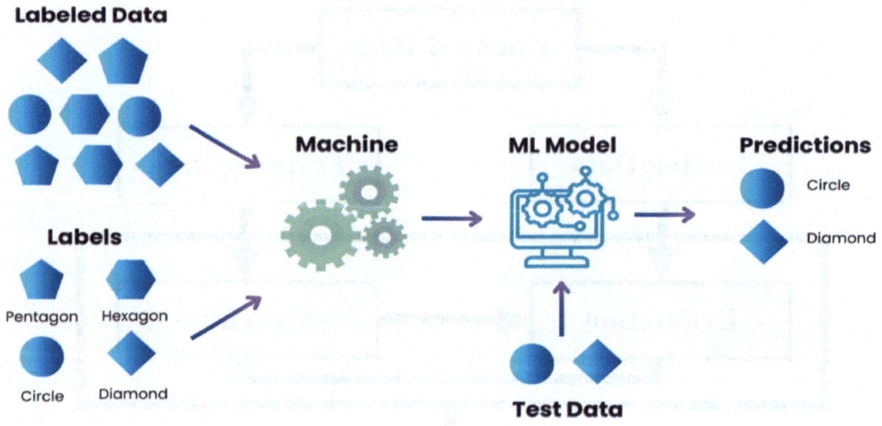

Fig. 3 Supervised machine learning process

2.2.1 Supervised Learning

A subset of ML and AI is supervised learning which is illustrated in Fig. 3. It is significant because it uses labeled data sets for training algorithms in order to sort out data or to forecast outcomes precisely. When we want to build a model, for example, that can identify images of shapes, engineers will feed a data set with labels that include a diverse range of shapes. Depending on the identified outcomes, the algorithm will determine which set of attributes corresponds to each shape. The model can be evaluated by presenting it with a portrait of a shape and asking it to identify the shape. If the model returns an improper response, we may keep training and changing its parameters with new instances to develop accuracy and diminish faults.

Supervised learning can be of many different kinds, such as linear regression, Naive Bayes, neural networks, logistic regression, SVM, K-nearest neighbors, random forests, and many more [3].

2.2.2 Unsupervised Learning

This kind of machine learning uses algorithms to evaluate and cluster unlabeled data sets. These algorithms identify several hidden patterns or data groups without human involvement. Unsupervised machine learning models, unlike supervised machine learning, are given unlabeled data and are allowed to discover insights and patterns without obvious guidance or instruction. Figure 4 demonstrates this type of machine learning process. The main functions of this method are association, clustering, and dimensionality reduction [4]. There exist three distinct categories of unsupervised learning methodologies: Clustering, Dimensionality, and Association. Clustering is the process of organizing data into meaningful groups. Dimensionality reduction is

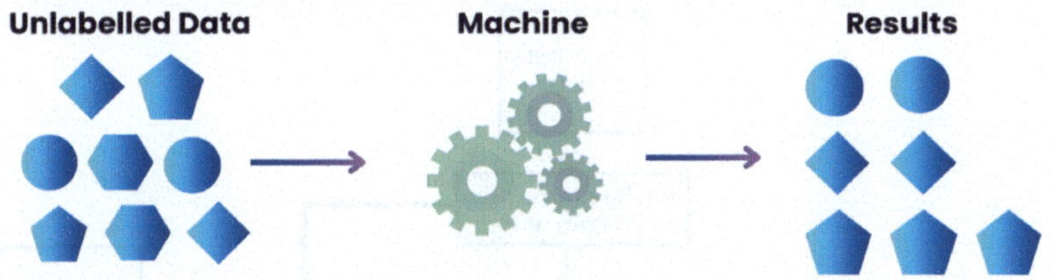

Fig. 4 Unsupervised machine learning process

the process of representing high-dimensional data with a lower number of dimensions. Association refers to the process of identifying the associations between variables within a large data set. Clustering refers to the process of grouping similar items or data points together based on their characteristics or attributes. Clustering occupies a prominent position. This technique uses a clustering algorithm to organize the data by identifying patterns.

One common real-world application of the approach is to group clients that share similar traits. Dimensionality reduction techniques aim to create a lower-dimensional representation of high-dimensional data. This technique is crucial because handling data with a large number of dimensions is challenging for both humans to read and machine learning algorithms to learn patterns from. This technique is frequently employed during the pre-processing stage of data, as intricate input material is condensed to its most valuable and informative components.

For example, we imagine that we have a large weather data set. An algorithm of the unsupervised learning will endure the data set to recognize patterns in the points. Currently, the most commonly used unsupervised learning algorithms are k-means clustering, fuzzy means, hierarchical clustering, and partial least squares [5].

2.2.3 Semi-Supervised Learning

Semi-supervised learning is one kind of ML that uses both unlabeled and labeled data to train AI models to classify and for regression problems. It consists of a few of labeled examples and a lot of unlabeled samples [6]. This method sits between supervised learning and unsupervised learning. They are especially useful in some cases when acquiring enough labeled data is excessively challenging or expensive, yet big volumes of unlabeled data are easy to obtain. In such states, neither fully supervised nor unsupervised learning methods are appropriate.

Transductive learning, co-training, and self-training are a few of the methods utilized in semi-supervised learning [7]. In this scenario, self-training employs a supervised learning algorithm-trained model on labeled data to forecast labels for unlabeled data. Once we add the most accurate forecasts to the set of training variables, we retrain the model using this enhanced data set. A machine learning pipeline intended for prediction and classification is depicted in Fig. 5. First, characteristics are retrieved and optimized from labeled data. K-means clustering is used to extract

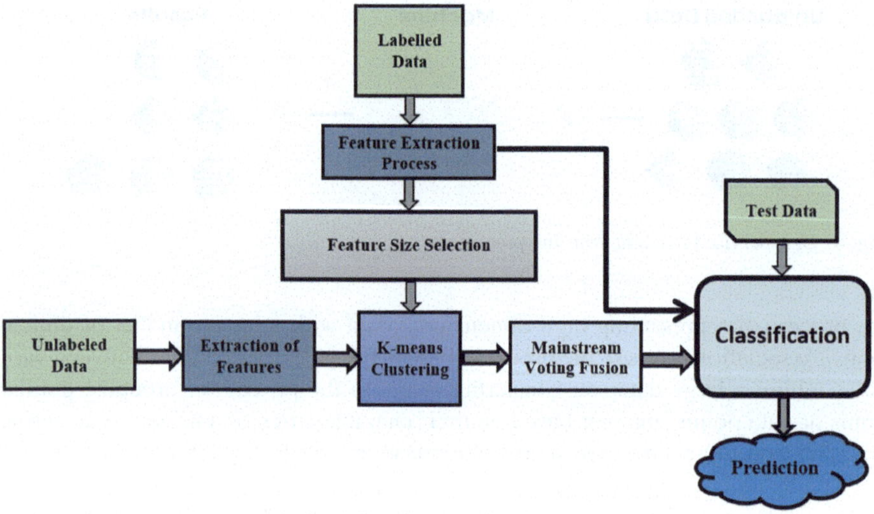

Fig. 5 Semi-supervised learning

features from unlabeled data concurrently. To improve accuracy, mainstream voting fusion is used to merge the clustering results. The classification model receives input from both the clustered unlabeled data and the processed labeled data. In order to ensure correct results, test data is finally identified and used to make predictions. This is done by using the trained model. For reliable prediction, our integrated method effectively handles both labeled and unlabeled data.

2.2.4 Reinforcement Learning (RL)

RL is a segment of ML that enables an AI-powered system (also known as an agent) to obtain knowledge via trial and error based on feedback from its previous actions. This feedback is either positive or negative, communicated as a reward or penalty, and its aim is to maximize the reward function. RL learns from its errors and provides AI that nearly resembles natural intelligence as is now achievable. Figure 6 demonstrates the total phenomena of the reinforcement learning process. In terms of learning approaches, RL is comparable to supervised learning in that it involves input–output mapping, but that is all they have in common.

Feedback directs the agent in supervised learning to the proper set of behaviors to implement. The agent determines how to do the assignment appropriately. Compared to unsupervised ML, RL has distinct objectives. The objective of unsupervised ML is to identify similarities and contrasts between data points. The purpose of RL is to develop the best action model to maximize the overall reward for the RL agent. In the absence of a training data set, the agent solves the RL issue using its own behaviors and environmental information.

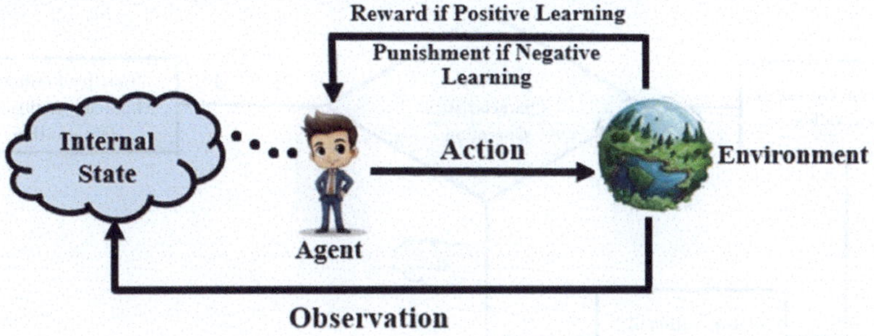

Fig. 6 Reinforcement learning

3 Evolution from Machine Learning to Artificial Intelligence

From its origins in machine learning (ML), an essential component that focuses on creating algorithms that allow computers to train from and take decisions depending on data, AI has seen substantial evolution. To create a framework for increasingly sophisticated AI systems, machine learning initially focused on identifying patterns and making data-driven predictions. The development of machine learning (ML) techniques, such as artificial neural networks and deep learning, allowed them to be more comparable to human cognitive procedures, which prepared the way for the expansion of AI applications. As a result of this development, artificial intelligence has been able to expand its applications beyond limited, specific tasks to include natural language processing, autonomous systems, and sophisticated problem-solving. Thus, the convergence of ML and AI has contributed to the creation of ever more intelligent systems that can carry out a broad range of tasks that are comparable to humans.

4 Application of ML, AI, and DL

4.1 Enhancement of Predictive Maintenance with Artificial Intelligence

A condition-based maintenance strategy guides the work of predictive maintenance (PdM). It includes maintaining a check on the tool's operational characteristics, evaluating operational data or performance, identifying potential problems, and addressing them before the equipment malfunctions [8]. Figure 7 illustrates the framework of the predictive maintenance process.

Facility maintenance and operations are characterized as the regular activities, functions, and manpower necessary to run and sustain a facility asset, ensuring that

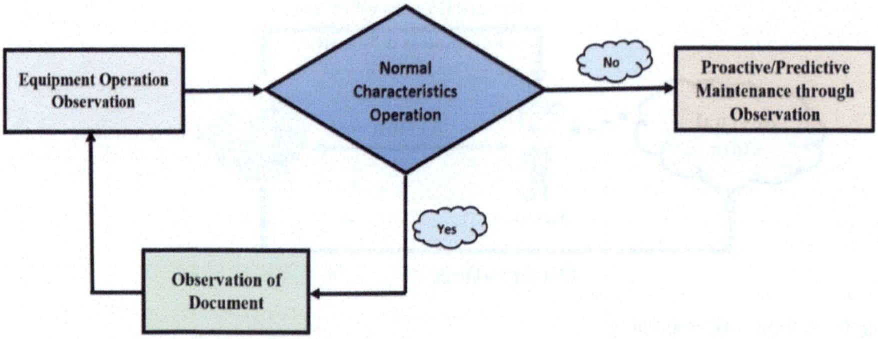

Fig. 7 Predictive maintenance framework

its purpose is available for the basic use and the functions are preserved over the life of the facility. Administrations, facilities, management experts, and patrons spend a lot of money each year to execute this duty. Scarce amount of facility operations is responsible for a major amount of the costs, which we can ignore. In this case, we may use AI in combination with prognostic maintenance in order to reduce operating and maintenance costs. The core concept of predictive maintenance is AI's ability to deliver unbiased investment and repair suggestions based on analyzed data set. A facility operations program cannot completely integrate AI owing to a scarce of data [9].

4.2 Applications of AI in Facility Management and Predictive Maintenance

AI enables real-time monitoring process and operational features of awareness of the facility. Unbiased and data-driven facility system maintenance, investment, and decisions. AI in the facility operations software collects, analyzes, and interprets data based on learnt features, then predicts based on the interpreted data.

4.3 Increase Facility System Online Time

Incorporating AI into the facility operations skill of a facility management program has the ability to boost facility system uptime. Predictive maintenance permits the surveillance and inspection of facility system operations for minor deviations, the resolution of issues before they become significant, and the smooth running of the facility [10]. Maintenance workers often plan and schedule equipment downtime during off-peak hours to avoid interrupting or disrupting existing operations, which is an advantage of predictive maintenance provided by artificial intelligence [11].

4.4 Smart Energy Management System

AI models estimate load profiles and energy consumption, as well as resource scheduling, to ensure efficient use of energy resources and consistent performance. AI is used in a variety of smart energy management applications, such as energy generation forecasting [5, 8, 9], demand forecasting [9, 12], demand side management, optimized energy storage operation, energy pricing prediction, weather phenomena related to energy predictions [9], energy theft detection, predictive maintenance and control [12], and building energy management [13]. Artificial intelligence models may be used in solar, geothermal, wind, hydro, ocean, hydrogen, bio, and hybrid energy [14]. The transformative power of AI provides several upsides to the energy industry. It reassembles the process of generation and delivery of energy. Figure 8 demonstrates the uses of AI in the field of energy sector, for energy management, and how the whole energy segment is digitalized.

AI models include wavelet neural networks (WNNs), artificial neural networks (ANNs), support vector machines (SVM), hybrids, decision trees, and ensembles. Hybrid ML models are often quicker, more precise, and simpler to use [9, 15]. Current

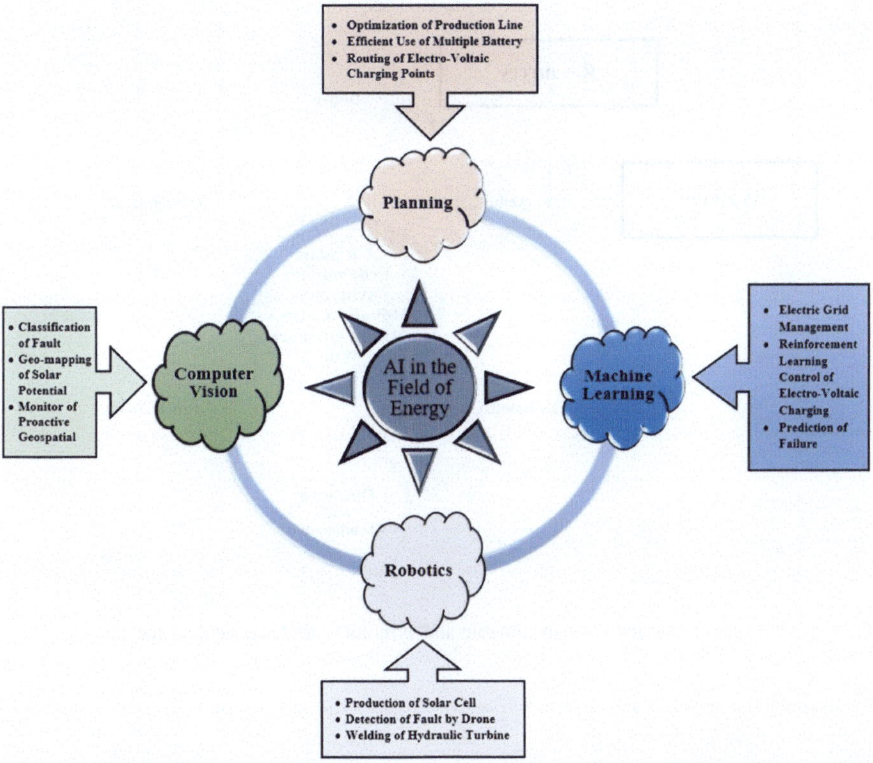

Fig. 8 Usage of AI in energy sector

and projected future energy management developments include the use of hybrid and multi-stage intelligent algorithms for energy forecasting and resource control [15].

4.5 *Robotics and AI (RAI) in Case of Offshore Wind Sector*

RAI are being utilized to remove lifetime service constraints and enhance beneficial and sustainable offshore wind energy production. Today, engineers employ RAI programs to help them accomplish short-range goals in operations and maintenance. In order to move forward, RAI can play an important role in the whole lifetime of offshore wind infrastructure, including surveys, planning, design, operational support, logistics, training, and discharging. The global need for clean and sustainable energy sources is a major driving force behind the progress of wind turbine design. Figure 9 highlights the economy circle of robotics and AI for a synergetic and sustainable offshore wind segment. Generally, wind farm operators use technical skills and innovative AI algorithms to guide wind turbine and wind farm design [16].

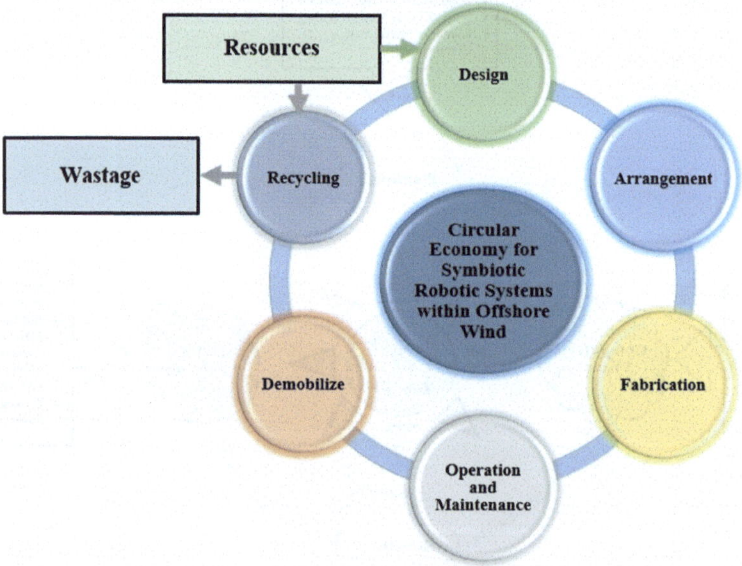

Fig. 9 RAI circular economy for a sustainable and symbiotic offshore wind sector

4.6 Application of AI and ML in Robotics

Machine learning and artificial intelligence are relatively recent developments. They are evolving, reaching new heights, and significantly enhancing their abilities. AI and machine learning can significantly improve robotics applications such as image recognition, navigation, prediction, and manufacturing. AI and machine learning have the potential to revolutionize robotics through enhanced image identification, navigation, anticipation, collaboration, and production ability. The autonomy of a robot in an environment is divided into three distinct groups: perception, planning, and execution. The basic objective of merging robots with AI is to achieve maximum autonomy via learning. It is reasonable to suppose that intelligence is defined by the ability to predict the future while arranging a task or engaging with the outside world.

AI plays a crucial role in perception in the field of robotics. Robots may use computer vision or integrated sensors to detect their environment. Over the past 10 years, computer systems have improved their vision and sensing capabilities. Not only is vision important for planning, but it also helps the robot develop a false feeling of self-awareness. This allows for interactions between the robot and other things in the same environment. This discipline is referred to as social or community robotics. It divides into two types: human–robot interfaces (HCI) and cognitive robotics. HCI seeks to improve the robotic perception of people in areas like activity understanding, emotion recognition, nonverbal communication, and environment navigation alongside humans [12, 17]. The ultimate goal of cognitive robotics is to enable robots to learn and acquire knowledge independently from complex levels of perception by imitation and experience [13, 18]. It tries to replicate how experience and use of sensor manage the human cognitive system, which governs the process of acquiring knowledge and comprehending [19]. Furthermore, there are approaches for cognitive robots that leverage curiosity and incentives to improve the quality and pace of information acquisition via learning [14].

Artificial intelligence has continued to break all past records and overcome various challenges that were unbearable less than ten years ago. These advancements, when combined, will continue to alter how we think about robotic intelligence in a range of new sectors. Furthermore, in March 2019, Xinhua News Agency that is controlled by the Chinese government has announced the introduction of their newest AI news presenter, a female-gendered scheme dubbed Xin Xiaomeng [10, 20]. According to AI Applications, enormous amounts of data must always be processed in order to explore and find space. AI and ML are the most effective ways to manage and handle such large amounts of data. After doing considerable research, scientists discovered a distant solar system with eight planets by using AI to comb through several years of data. Kepler telescope was used to record the data. Self-driving cars have long been a buzzword in the field of artificial intelligence. The arrival of self-driving cars will surely revolutionize the transportation sector. Some businesses, such as Waymo, have tested the first AI-driven public ride-hailing service [20, 21].

4.7 AI and Robotics in Healthcare

AI is becoming progressively capable of performing human-like tasks more effectively, swiftly, and affordably. Robotics and AI have enormous potentiality in healthcare. They are becoming increasingly integrated into our healthcare ecosystem, just as they are in our daily lives. Eight ways to showcase the transformation of healthcare system via AI and robotics have been highlighted in Fig. 10. AI is helping people to stay healthy and fit through health applications. Besides, the specialists can detect and diagnose fast to make an early decision, to provide treatment, and the end of life has been approached.

Elderly people are more comfortable with home assistance robots because they are more open to mobile devices. Robots can perform household tasks and basic daily operations. Smart houses, service robots, and intelligent agents help senior citizens with daily tasks. This allows them to live freely and without invasive activities for as long as possible. IoT, cloud-enabled services, and AI will be examples of enabling technology [22]. The COVID-19 pandemic and the Ebola outbreak have revealed numerous applications of robotics [23]. As the use of robotic surgery grows, more surgeons are beginning their careers with it. Furthermore, an increasing number of residencies include robotic surgery in their basic training. Competency-based robotics training programs have been created. We included them in the training of

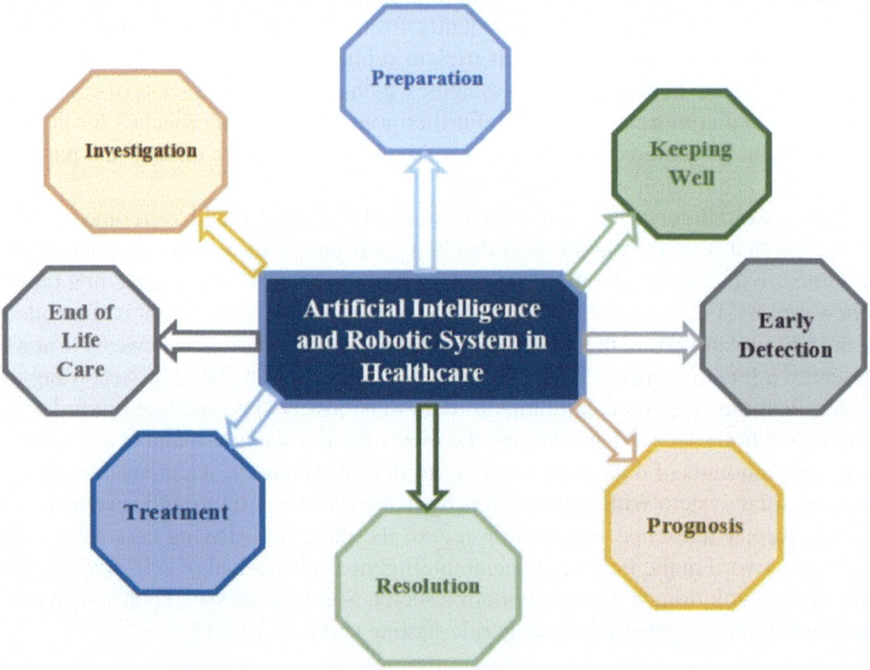

Fig. 10 AI and robotics in healthcare system

surgical oncology fellows. The findings revealed that the fellows could effectively accomplish sophisticated robotic gastrointestinal operations following training and utilize their newly acquired skill set after graduation.

According to studies comparing open versus robot-assisted laparoscopic surgery (RALP) for prostate cancer, patients who chose the latter had better mean educational levels, despite an approximately 30% increase in operation expenses. In terms of consequences, researchers connected the open technique to a greater probability of thromboembolism, while RALP was related to a lower rate of transfusion and a shorter time. A mathematical model was proposed as a result of research on the application of robots in surgery. SARA, a system for socially assistive robotics (SAR), is currently being developed. The system's goal is to give an automatic health monitoring system and support to elderly persons who have modest cognitive impairments. SARA's challenges include security, privacy, reliability, data interoperability, IoT network management, and device. The system supports robot activity (RAS). Robotic surgery can reduce surgery time while improving surgical workflow efficiency. Robots help to adjust workflow and execute procedures on various body regions. Robotic surgery advances image processing using neural networks, as well as the digitalization of radiography and pathology. It allows for precise motions during bi-manual operation, with articulated arms and a 3D magnified perspective. However, the effectiveness of robotic surgery is dependent on the quality of the data given. Despite this, the healthcare business still has privacy issues. Surgeons can get useful insights from the information communicated and obtained during surgery. Standard diagnostic techniques include brachytherapy, breast biopsies, and prostate biopsies. We can perform these treatments using robotic percutaneous needle applications guided by ultrasonography [24]. We recommend the use of robotic machinery since these scenarios necessitate a certain degree of hand dexterity. The technology's use enhances the accuracy of doubtful tissue identification. It allows for the removal of a biopsy through small lesions. Similarly, human physicians can use robots in MRI-guided treatment to mimic the physical two-point grasping motion. A robotic system can carry out small-scale manipulation tasks that involve human judgment, sensory integration, and hand–eye coordination. We call it a steady-hand micromanipulation system [13]. This allows the operator's hand to handle surgical tools alongside a specially designed robot arm.

Medical imaging has become an essential component of healthcare over the last few decades. Imaging or image processing is widely used for illness treatment, diagnosis, verification, and rehabilitation. Doctors examine digital medical pictures from several sources, including retinal photography, mammography, morphology slides, histology, ceroscopy, computed tomography (CT), ultrasound (US), magnetic resonance imaging (MRI), and positron emission tomography (PET) scans. Data about diagnostic imaging modalities and their most common applications is available. Scientists aim to apply methods from deep learning to a number of healthcare jobs. Some of these activities include classification, lesion detection, area of interest (ROI) segmentation, registration, image generation, image quality enhancement, and integrating image data with clinical reports to provide a written report based on image analysis. In this example, we focus on applications and different approaches to addressing the stated issues in various anatomical locations, such as the brain, eye, chest, breast, heart, and abdomen [25–28].

MRI has a high spatial resolution and is essential for analyzing brain anatomy, whereas AI models for neuroscience studies mostly concentrate on MRI. Other modalities with inferior contrast decisions, like CT and resting-state fMRI, have fewer models. As a result, radiologists prefer T1W MRI over other modalities when performing 3D sequences. However, the use of AI analysis and automatic analysis of lesions would greatly benefit patients and experts.

The primary goals of thyroid gland research are to detect malignant illness in thyroid nodes and forecast lymph metastases. Researchers employ artificial intelligence to improve the TI-RADS risk rating for thyroid nodules. The primary areas of study on the thyroid gland are the detection of malignant illness in thyroid nodes and the prediction of lymph metastases. Researchers employ artificial intelligence to improve the TI-RADS risk rating for thyroid nodules. Swallowing creates movement artifacts, so MRI is ineffective, and US has limited sensitivity for identifying lymph node invasion. Some researchers created an automated approach for recognizing thyroid and neck tissues in ultrasonography using a deep learning architecture [29].

The primary drawback of the lymph node research was that, like previous AI studies, the authors regarded each node as a distinct instance in order to expand the size of the training data set. With this strategy, the model may get over-trained to predict specific tumor types that misestimate more frequently. Because ophthalmology uses diagnostic images extensively, the corresponding AI models require preoperative planning and diagnosis. The researchers are now exploring the potentiality of AI in dry eye disease screening to identify the challenges and future research directions [28]. Machine learning mainly enhances ophthalmology by improving the accessibility associated with the diagnosis at a very low cost [30, 31].

X-ray mammography is one common subject of AI research in computer vision medicine. There are two main reasons for this. First, screening for breast cancer often involves the use of breast X-rays. Secondly, this approach has a low detection rate for soft tissue disease. Researchers frequently make the approach more sensitive to the disease's early stages. First, screening for breast cancer often involves the use of breast X-rays. Secondly, this approach has a low detection rate for soft tissue disease. Researchers typically make the approach more sensitive to early illness stages. Despite the effectiveness of US as a breast cancer screening tool, the primary purpose of US-based AI solutions is to predict the condition of lymph nodes or detect their metastases. Because it's a useful diagnostic tool, new research uses MRI mammography data to feed machine learning algorithms because it's a useful diagnostic tool as well.

The amount of research on using CT to identify lung cancer is increasing. Figure 11 demonstrates the forecast of lung cancer through artificial intelligence. Besides, AI methods enable automatic pulmonary node detection and differentiation between benign and malignant nodes. One common shortcoming of the systems is that they lack the ability to differentiate between pathologies that warrant differential diagnosis and lung cancer. For example, we were unable to develop an AI method for sarcoidosis detection. If we were to build one, it would greatly benefit pulmonology. Using AI

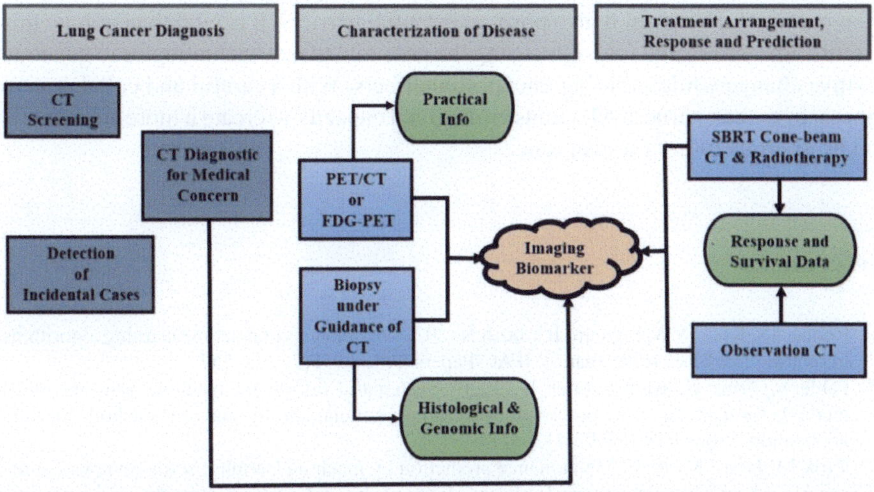

Fig. 11 Lung cancer prognosis by using AI

to diagnose pulmonary tuberculosis is one area of application. One common issue is that data scientists mostly work with 2D images (CXR) due to data availability, despite their low informative value. Chest CT scans, which are far more sensitive and frequently performed on patients, inspire a few models. Machine learning techniques for detecting COVID-19 emerged shortly after the outbreak. By using lung CT appearance to deliver sensitive diagnoses, they assist medical professionals in differentiating between community-acquired pneumonia and COVID-19.

Moreover, artificial intelligence provides several solutions for the abdominal cavity. The abdominal cavity radiology includes fat tissue and models of segmentation to evaluate the volume of inner organs and classify models for a radiation response prediction qualitatively [13].

5 Conclusion

To conclude, this chapter provides an informative investigation into the fields of AI and ML. The fundamental concepts that underpin these technologies have been investigated here. The range of the investigated technology ranged from the fundamental principles of learning algorithms to the intricate designs of neural networks. Through case studies and examples, we have seen the revolutionary impact of AI in a variety of disciplines, including healthcare, finance, and transportation. However, as we delight at the ability of a robotic body to develop and adjust, we must equally consider the moral and social impacts of this rapid expansion. Questions of prejudice, privacy, and job displacement loom big and need serious study and ethical management by everyone involved. Moving into the future, the next phase of AI and ML has great potential. However, it also poses substantial barriers. It is vital to negotiate this terrain with caution and foresight, using the potential of new technologies to generate positive change while avoiding unexpected effects. With a careful and collaborative approach, we can harness AI's transformational capacity to create a more sustainable and prosperous future for everyone.

References

1. Roque AS, Krebs VW, Figueiro IC, Jazdi N (2022) An analysis of machine learning algorithms in rotating machines maintenance. IFAC-Papers OnLine 55(2):252–257
2. Fahle S, Prinz C, Kuhlenkötter B (2020) Systematic review on machine learning (ML) methods for manufacturing processes–identifying artificial intelligence (AI) methods for field application. Procedia CIRP 93:413–418
3. Tarik M, Jebari K (2020) Maintenance prediction by machine learning: study review of some supervised learning algorithms. In: Proceedings of the 2nd African international conference on industrial engineering and operations management. IEOM Society International, Harare, Zimbabwe
4. Samuel OW, Asogbon MG, Geng Y, Al-Timemy AH, Pirbhulal S, Ji N, Li G (2019) Intelligent EMG pattern recognition control method for upper-limb multifunctional prostheses: advances, current challenges, and future prospects. Ieee Access 7:10150–10165
5. Bengio Y, Courville A, Vincent P (2013) Representation learning: a review and new perspectives. IEEE Trans Pattern Anal Mach Intell 35(8):1798–1828
6. Zhang D, Lee WS (2004) Learning to integrate web taxonomies. J Web Semant 2(2):131–151
7. Wang Z, Li T, Zheng JQ, Huang B (2022) When cnn meet with vit: towards semi-supervised learning for multi-class medical image semantic segmentation. In: European conference on computer vision, pp 424–441. Springer Nature Switzerland, Cham
8. Lei Y, Yang B, Jiang X, Jia F, Li N, Nandi AK (2020) Applications of machine learning to machine fault diagnosis: a review and roadmap. Mech Syst Signal Process 138:106587
9. Theissler A, Pérez-Velázquez J, Kettelgerdes M, Elger G (2021) Predictive maintenance enabled by machine learning: use cases and challenges in the automotive industry. Reliab Eng Syst Saf 215:107864
10. Paleczek A, Grochala D, Rydosz A (2021) Artificial breath classification using XGBoost algorithm for diabetes detection. Sensors 21(12):4187
11. Scaife AD (2023) Improve predictive maintenance through the application of artificial intelligence: a systematic review. Results Eng 101645
12. Burduk A, Chlebus E, Nowakowski T, Tubis A (2018) Advances in intelligent systems and computing intelligent systems in production engineering and maintenance, vol 835. Available: http://www.springer.com/series/11156
13. Habuza T, Navaz AN, Hashim F, Alnajjar F, Zaki N, Serhani MA, Statsenko Y (2021) AI applications in robotics, diagnostic image analysis and precision medicine: current limitations, future trends, guidelines on CAD systems for medicine. Inform Med Unlocked 24:100596
14. Oudeyer PY, Smith LB (2016) How evolution may work through curiosity-driven developmental process. Top Cogn Sci 8(2):492–502
15. Li J, Herdem MS, Nathwani J, Wen JZ (2023) Methods and applications for artificial intelligence, Big Data, Internet of Things, and Blockchain in smart energy management. Energy AI 11:100208
16. Mitchell D, Blanche J, Harper S, Lim T, Gupta R, Zaki O, Flynn D (2022) A review: challenges and opportunities for artificial intelligence and robotics in the offshore wind sector. Energy AI 8:100146

17. Ran Y, Zhou X, Lin P, Wen Y, Deng R (2019) A survey of predictive maintenance: systems, purposes and approaches. arXiv preprint arXiv:1912.07383.
18. Soori M, Arezoo B, Dastres R (2023) Artificial intelligence, machine learning and deep learning in advanced robotics, a review. Cogn Robot
19. IEEE Staff and IEEE Staff, 2010 international conference on computational intelligence and software engineering
20. Mihret ET (2020) Robotics and artificial intelligence. Int J Artif Intell Mach Learn (IJAIML) 10(2):57–78
21. Bonaccorsi M, Fiorini L, Cavallo F, Saffiotti A, Dario P (2016) A cloud robotics solution to improve social assistive robots for active and healthy aging. Int J Soc Robot 8:393–408
22. Dubey AK, Kumar A, Kumar SR, Gayathri N, Das P (eds) (2021) AI and IoT-based intelligent automation in robotics. John Wiley & Sons
23. Bhaskar S, Bradley S, Sakhamuri S, Moguilner S, Chattu VK, Pandya S, Banach M (2020) Designing futuristic telemedicine using artificial intelligence and robotics in the COVID-19 era. Front Public Health 8:556789
24. Mao Y, Zhang H, Xu N, Zhang B, Gou F, Zhu JK (2013) Application of the CRISPR–Cas system for efficient genome engineering in plants. Mol Plant 6(6):2008–2011
25. Mavridis N (2015) A review of verbal and non-verbal human–robot interactive communication. Robot Auton Syst 63:22–35
26. Xu W, Huang J, Wang Y, Tao C, Cheng L (2015) Reinforcement learning-based shared control for walking-aid robot and its experimental verification. Adv Robot 29(22):1463–1481
27. Horst C, O'Shea R, Goh V (2022) Lung cancer characterization and prognosis: the role of artificial intelligence. Artificial intelligence in cardiothoracic imaging. Springer International Publishing, Cham, pp 471–481
28. Deutsch I, Erel H, Paz M, Hoffman G, Zuckerman O (2019) Home robotic devices for older adults: opportunities and concerns. Comput Hum Behav 98:122–133
29. Ma L, Tan G, Luo H, Liao Q, Li S, Li K (2022) A novel deep learning framework for automatic recognition of thyroid gland and tissues of neck in ultrasound image. IEEE Trans Circuits Syst Video Technol 32(9):6113–6124
30. Armstrong GW, Lorch AC (2020) A (eye): a review of current applications of artificial intelligence and machine learning in ophthalmology. Int Ophthalmol Clin 60(1):57–71
31. Wang MH, Xing L, Pan Y, Gu F, Fang J, Yu X, Liu W (2024) AI-based advanced approaches and dry eye disease detection based on multi-source evidence: cases, applications, issues, and future directions. Big Data Min Anal 7(2):445–484

Emerging Applications of Mechatronics

Rezwan U. S. Saleheen, Ahmed Farhan, Nusrat Zahan Ramesha, Rumana Tasnim, Md. Tarek Ur Rahman Erin, and Sumaiya Shahria

Abstract Mechatronics engineering is a creative and dynamic field that focuses on cutting-edge technology for a variety of applications including high-speed manufacturing systems used in many contemporary companies. Modern manufacturing incorporates sophisticated technology and cutting-edge software, tools, and goods that improve quality of life utilizing intelligent sensors and parameter controllers. Mechatronics has enhanced efficiency and the quality of the products while decreasing technical faults in smart autonomous manufacturing. A detailed overview of the recent mechatronics research with applications in manufacturing, agriculture, industrial automation, robotics, biomedical and assistive technology, human–machine interface, unmanned vehicles, energy, aerospace, and transportation is provided in this chapter. Production has significantly increased as a result of Industry 4.0, digitization, and artificial intelligence integration in mechatronics. A significant factor in the change of the automobile industry has been the development of mechatronics in electric and driverless vehicles. Numerous research and development initiatives have the potential to advance the use of mechatronics in line with forthcoming findings. Technology based on mechatronics focuses on practically all facets of intelligent industrial techniques and is useful for every area of a high-quality way of life for almost everyone.

Keywords Technology · IR 4.0 · Autonomous manufacturing · Artificial intelligence · Sensors

R. U. S. Saleheen (✉) · A. Farhan · N. Z. Ramesha · R. Tasnim
Department of Mechatronics Engineering, World University of Bangladesh, Dhaka, Bangladesh
e-mail: saleheen1@mte.wub.edu.bd

Md. T. U. R. Erin
Department of Mechanical Engineering, Textile Engineering College, Chattagram, Bangladesh

S. Shahria
Department of Mechatronics Engineering, Khwaja Yunus Ali University, Enayetpur, Bangladesh

M. M. Rahman et al. (eds.), *Mechatronics*, Emerging Trends in Mechatronics,
https://doi.org/10.1007/978-981-97-7117-2_7

1 Introduction

New approaches to the production processes have been introduced by the expanding technological advancements of complex systems and the increasing demands placed on their productivity, such as manufacturing equipment. A smart mechatronics machine including several adaptive controls may make judgments regarding the process in real time. The multidisciplinary engineering field of mechatronics technology offers a sophisticated skill set for managing production process control and quality control. Studying mechatronics technology is essential to modern industrialization, including the production of food, electronics, electrical systems, and automobiles (Fig. 7.1).

Enhanced opportunities to assess and fine-tune the automated system's overall dynamic motion performance early in the manufacturing process. The increasing impact of inter-component interactions on possible machine dynamics, precision, and component quality is another effect. Throughout the design phase and the synchronization of feed drives and frame structure components, numerous factors must be taken into account, including the machining process, the influence of the control system, and the interconnections between machine components. The future of products will see an increase in complexity and the integration of mechanical and electrical operations, along with a growing tendency toward embedded systems and ubiquitous computing. Global supply chains have caused a rapid evolution in the manner that businesses create and manufacture their products. Subsystems and components will increasingly be sourced from suppliers around the globe, and subassembly and final assembly activities will require them to work together flawlessly. If manufacturing facilities are added and removed to fulfill contract obligations and keep import taxes low on new foreign contracts, much more trouble will result. More

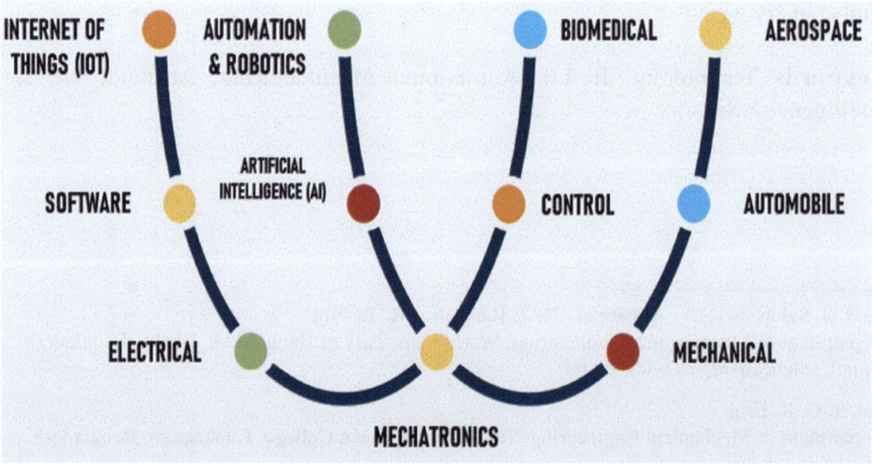

Fig. 7.1 Trends of mechatronics engineering

frequent switching out of suppliers and components from designs is also necessary to address the challenges posed by global supply chains and cost-cutting constraints. The increased flexibility of manufacturing will require it to swiftly adapt to new sites, component suppliers, and design modifications. Modern mechatronic technologies will be needed for each of these factors—from manufacturing design to ongoing process monitoring and modification—to ensure high-quality manufacture. They will be crucial for both profitability and excellent quality, claim [1].

Primarily, mechatronics is an approach to designing electromechanical products as efficiently as feasible. More than forty years ago, engineer Tetsuro Mori created the term "Mechatronics" by fusing the words "mechanical" and "electronic" to refer to the electronic control systems Yukawa Electric Corporation was creating in 1969 for the mechanical production equipment. The participation of these domains throughout the design process is the main component of mechatronics. Mechatronics creates the perfect environment to increase synergy by replicating transdisciplinary concepts and methods. This acts as a stimulant for the creation of original answers to technically difficult problems. One of the main characteristics of mechatronic systems and gadgets is their innate intelligence. Precise mechanical and electrical engineering, real-time programming included in the design process, and other elements combine to provide this intelligence. Synergy is the phenomenon wherein an entire performs better than the sum of its parts when certain circumstances are satisfied. Products that are mechatronic display performance attributes that were previously hard to achieve apart from the cooperative relationship. Recent research articles have demonstrated several mechatronic applications on robotic arms, helicopters, and micromotion [2–6]. Future product development processes will need to employ increasingly sophisticated techniques due to the growing demands on machine tool efficiency and technical complexity. Integrated technologies for virtual prototyping are employed early in the development process, beginning with the mechatronics design phase, to examine open architectural challenges, interactive modeling, and virtual prototyping. Intelligent gadgets for online and real-time monitoring, such as process control and diagnosis, also have an impact on mechatronics. Recently, mechatronics has made significant progress in smart manufacturing and its potential to enhance and adapt conventional designs [7].

2 Trends of Mechatronics and IR 4.0

Mechatronics technology is strongly tied to the recent spread of Industry 4.0. Mechatronics technology already incorporates essential features of sophisticated technology, such as robotics, the Internet of Things, big data, and data analytics, with sensor technology. Industry 4.0-enhanced automated manufacturing facilities are the direction of the industry trend. Automation technician, robotic technician, mechatronics technician, automated equipment maintenance technician, etc. are some possible careers related to mechatronics technology. The most recent digital electronics' intelligence and capabilities are assisting in elevating mechatronics to a new level by creating smarter systems that enable users to "plug-and-produce, perform, and proceed." This is partly due to the reason that manufacturers are requesting production equipment that is simple to operate and versatile. Simplifying the engineering effort for systems that may swiftly shift to create different versions on the same production line is intended to achieve a faster time to market. Mechatronic component specification, procurement, and commissioning are all streamlined by engineers through the use of online tools that provide fully functional, intelligent systems that require little to no machine programming to work "out of the box." Meanwhile, transparent production processes, real-time condition monitoring, and predictive maintenance features are essential for next-generation mechatronics. Ultimately, the demands of industry and the deployment of automated machinery suggest that end users desire easy connectivity with machine-level and plant-wide production management systems, automated production data tracking, and precise control and execution of motion sequences.

In many processing stations, the typical efficient processing mode can be subverted by combining non-standard mechanical structure design with an increasing quantity of control technology that is produced and practiced as a result of the growth of industrial control software technology. The development of artificial intelligence has meant that the two work well together, which is good news for many manufacturing companies. AI-based technologies and mechatronics work quite well together. Technological innovation in mechatronics is necessary to fulfill the unique demands of the present era of mechanical and electrical engineering construction. Technological advancements in microelectronics have revitalized mechatronics and offered it hope for the future. The nation's mechanical and electrical engineering sectors are expanding, and with them, so are the requirements for building technology, which includes mechatronics, or the application of microelectronics technology. This will push the mechanical and electrical engineering fields into new development areas and facilitate the seamless integration of mechatronics. Moreover, the common operation of mechatronics technologies will facilitate their regular operation [8].

3 Programmable Controllers: A Revolutionized Manufacturing Process

Mechatronics technology has reached a more developed stage from the middle of the twentieth century, despite the fact that previous microelectronics technology can still have some influence on its development. Microelectronics technology is unable to keep up with the real application fields of mechatronics technology and address every issue that arises as the new era develops. A programmable controller's main objective is to automate processes by utilizing its programmable features. The technology of controllers that combine electrical and mechanical parts is always getting better because of all the useful ways that new controllers can be used in production. Following the creation of this technology, the state set a number of management standards to ensure the proper oversight of the latest electromechanical controller that has been in use currently. Automated mechatronics are made possible by mechatronics technology, which gives mechatronics intelligence. Information is provided via the remote control interface to each receiver because of the rapid screening and aggregation capabilities of the new information processing technology. A new era of mechatronic systems is being ushered in by modern mechatronics technology, which is primarily concentrated on its application in research and development.

The ongoing progress in science and technology is leading to an increase in the scale of mechatronics. Controllable programmers are unable to keep up with the advancement of these new mechatronics businesses; therefore, they must move closer to intelligence in order to better support contemporary mechatronics businesses. The development of laser optoelectronics technology, which increases the bar for mechatronic intelligence, serves as evidence of this. A foundation for optoelectronics may be programmed manipulation. By guaranteeing a high current level, this would ensure that the intelligent evolution of mechatronics can be accomplished more quickly and effectively with the use of laser optoelectronics technology. With the use of laser technology, photonics can expedite the goal of intelligent advancement and facilitate mechanical integration. The mechatronics industry and the mechanical and electrical sectors of society should always be up to date on the most recent developments in science and technology. The field of Mechatronics and the latest advancements in science and technology have been skillfully combined with the dynamics development trend and the way individual enterprises are changing within the social development process. Mechatronics technology is keeping pace with this trend by prioritizing mechatronics research and meticulously developing prerequisites. According to the advantages of growing its own interdisciplinary businesses, it should prioritize research into technology related to the integration of electrical and mechanical to develop cutting-edge, highly efficient, and high-quality engineering devices. Such intelligent development ensures that its electromechanical enterprises can always be at the forefront of global development [9].

4 Emerging Applications of Mechatronics Engineering

4.1 Mechatronics in Additive Manufacturing System

Additive manufacturing, popularly referred to as 3D printing, is a ground-breaking mechatronics engineering application that creates three-dimensional things [10]. Unlike traditional manufacturing techniques, which often include subtractive steps like cutting or shaping materials, additive manufacturing starts with a computer design and creates the final object instantly by carefully adding material. With the use of this technology, intricate and complicated geometries that would be difficult or impossible to create with conventional methods can be created. The degree of design freedom that additive manufacturing provides sets it apart from conventional methods. With additive manufacturing, designers may create complicated structures, internal channels, and organic shapes more easily and freely than they can with traditional production procedures. This facilitates the production of highly customized and unique goods that meet specific requirements or individual tastes. Another unique aspect is the reduction of waste and resource use. Because raw materials must be chopped, machined, or shaped during traditional manufacturing processes, there is frequently significant material waste. Additive manufacturing, in contrast, uses just the material required for the final result to build items layer by layer. This lowers waste production and may result in lowering costs when it comes to the use of raw materials. Moreover, the product development cycle is shortened and quick prototyping is made possible via additive manufacturing. Prototypes and new product releases may be delayed by the lengthy setup, tooling, and production procedures associated with traditional manufacturing methods. However, additive manufacturing has the advantage of rapid design iterations and on-demand production, accelerating time to market and product development. Because digital design files and computer-controlled procedures are used in additive manufacturing to make real products, it is considered a digital technology.

Digital data forms the foundation of the entire process, from the first design to the finished output. Using computer-aided design (CAD) software, designers produce three-dimensional (3D) models that are subsequently translated into machine-readable instructions that direct the additive manufacturing devices. Rapid prototyping, intricate geometry, and personalization are made possible by its digital nature. Thus, one of the main forces behind the Fourth Industrial Revolution is additive manufacturing [11]. It bridges the gap between the real and digital worlds and is a major enabler of the digital transformation in manufacturing. Businesses can increase their production processes' efficiency, flexibility, and reactivity by incorporating additive manufacturing. It makes on-demand production possible, which lowers the cost of logistics and inventories. Moreover, additive printing can be combined with other digital technologies such as data analytics, artificial intelligence, and the Internet of Things (IoT) to build "smart factories" that maximize productivity, allow for predictive maintenance, and make product customization easier. The field of additive manufacturing holds significant promise for revolutionizing the corporate and economic spheres. Its potential to lower costs related to conventional production procedures is one of its main features. This is so that more effective production processes may be achieved, material waste can be decreased, and tooling can be eliminated due to additive manufacturing. Businesses can benefit greatly from these cost savings by becoming more competitive, being able to offer more inexpensive products, and possibly even reshoring manufacturing operations. More customization and design freedom are also made possible by additive manufacturing. This creates new avenues for business, particularly with customized goods.

Businesses are adopting additive manufacturing more and more to create personalized products, particularly in the sports and healthcare sectors. This capacity helps businesses stand out in the marketplace in addition to improving client happiness. Businesses can develop distinctive product offers, reach specialized customers, and position themselves as leaders in their respective fields by utilizing additive manufacturing. The way additive manufacturing is affecting supply chains is another way it demonstrates its disruptive potential. Conventional manufacturing frequently depends on drawn-out supply chains that include numerous suppliers and logistical processes. However, by facilitating on-demand and local production, additive manufacturing can simplify the supply chain. As a result, lead times are shortened, extra inventory is not needed, and the production system is made more flexible and responsive. AM has the ability to spur innovation, upend established business paradigms, and change entire industries. It is anticipated that technology will have a major impact on businesses and the economy as it develops and becomes more widely available. More and more materials are accessible with more and more qualities; certain parts produced using additive manufacturing (AM) now match or even exceed the quality of parts produced through traditional methods. Although technology is still in its early stages of development, it has already made great strides and is widely used across a range of industries. In addition to being widely employed for industrial production of end parts in the aerospace, medical, power and energy, and consumer markets (architecture, footwear, athletic equipment, and eyeglasses) additive manufacturing (AM) technology also speeds up product creation through prototyping. Furthermore, there is still a lot of unrealized potential in the automotive, fashion, food, and electronics industries. As the technology advances and encompasses new materials, printing technologies, and applications, it is anticipated to yield substantial development prospects for the manufacturing sector. Additive manufacturing has already shown its revolutionary potential by making it possible to create intricate geometries, cut lead times, and even customize goods as needed. As new materials, improved printing methods, and creative uses keep popping up, more research is needed to fully comprehend and utilize this technology. AM has the capacity to boost competitiveness and economic growth in the future. Additive manufacturing can also aid in addressing important problems with sustainability and environmental effects. With the growing awareness of sustainable practices, additive manufacturing presents an opportunity to minimize waste, maximize material utilization, and facilitate localized production.

4.2 Mechatronics in Industrial Automation

The need to support greater production and efficiency as well as technology improvements are driving a rapid advancement in industrial automation. The advancements in industrial automation and how they affect various industries highlight how important it is for businesses to identify and understand these trends in order to stay competitive in the dynamic business world. Significant progress has been made in industrial automation research recently. A few of them have made noteworthy contributions to improving productivity in industrial sectors by improving accuracy and efficiency. First of all, the combination of machine learning (ML) and artificial intelligence

(AI) techniques is transforming traditional automation systems and enabling intelligent decision-making, adaptive control, and optimized processes. Secondly, the Internet of Things (IoT) is revolutionizing industrial automation by tying together machines, sensors, and other devices. Predictive maintenance, real-time data transfer, and enhanced operational visibility result from this. Furthermore, advancements in robotics and automation are driving the development of autonomous systems, cooperative human–robot interaction, and the automation of challenging tasks. Additionally, big data analytics is essential to industrial automation because it can extract meaningful insights from massive data sets, facilitating better resource allocation, process optimization, and predictive maintenance. After careful consideration and debate, these implementations have an impact on the labor markets, companies, and sectors related to these emerging trends. Studies are being conducted to look at the implications for future industrial automation systems to identify commonalities, conflicts, and potential areas of overlap between these trends. Industry decisions about personnel planning, process optimization, and technology adoption can be made with knowledge of these new trends, which will eventually spur innovation and expansion [12].

In the manufacturing industry, automation is the use of machinery and equipment to automate and simplify different production processes and systems. The principal aim is to optimize overall efficiency through the concurrent expansion of production capacity and cost reduction. Assigning tasks that were formerly handled by humans to machines has been more synonymous with automation in recent years. It includes putting into practice electromechanical systems that are programmed to carry out various tasks. The topic of industrial automation is quickly developing and includes a range of technologies and trends that are intended to improve industrial operations and boost productivity. The industrial automation environment has seen significant change in the last few years due to a number of rising trends. These trends include developments in fields like data analytics, cloud computing, robots, artificial intelligence, and the Internet of Things. Industries can increase their levels of productivity, adaptability, and scalability by utilizing these technologies, which will ultimately boost overall performance.

Mechatronics design now requires a system-level approach because of the tremendous growth in device and system complexity. This method streamlines testing, integrates mechanical and control design, and allows for algorithm reuse inside the final embedded delivery framework. Investments in industrial machinery, renewable energy, medical research, and life sciences are what are driving this development. Engineers may produce virtual prototypes more quickly by dividing the design process into simultaneous threads. This reduces development time and costs and fosters teamwork among design teams. Device vendors such as National Instruments have invested in encouraging the use of built-in design instruments in mechatronics design processes. With the help of these interfaces, developers can establish a smooth path to embedded hardware and create virtualized versions of their systems, all while facilitating seamless communication across a multitude of devices. Embedded hardware is required for algorithm development and implementation. Developing a component-level tailored control system to install on embedded commercial-off-the-shelf (COTS) systems can help engineers save money and expedite production time during deployment. Before developing custom designs that are as economical as feasible, "system-ready" subsystems are usually employed in delivery platforms or prototypes to assess client approval. In recent years, there has been an increase in the popularity of customized embedded systems, which provide the speedy deployment of commercial off-the-shelf hardware along with the flexibility of a custom configuration. It has been seven years since National Instruments initially introduced the RIO (reconfigurable I/O) platform. It has since advanced and is now used in system-level solutions for industrial applications as well as durable embedded electronics. In the future, software will be crucial to mechatronics and embedded systems because it abstracts the complexity of cutting-edge technology and provides a system-level perspective on many design considerations. Three major developments in mechatronics for the future are the "ghost of computers," "the hidden life of data," and "the future of terrifying" [13].

4.3 Mechatronics in Health Care

Several areas of health care are revolutionized by mechatronics engineering, including patient care, diagnosis, therapy, and the creation of medical devices. Mechatronic-powered wearable medical devices are essential to contemporary health care because they allow ongoing patient engagement, early health issue diagnosis, tailored treatment plans, and continuous monitoring.

Researchers' interest in creating a healthcare system to track senior citizens' health-related activities has grown over the past few years [14–17]. Mechatronic systems allow remote patient and vital signs monitoring through wearable devices, home monitoring systems, and telehealth platforms. These systems gather, transfer, and analyze patient data in real time, enabling the early identification of health problems and prompt therapies by utilizing sensors, wireless connectivity, and data processing algorithms. Vital indicators including heart rate, blood pressure, temperature, and breathing rate can be continuously monitored via wearable technology that has built-in sensors. Real-time insights about a patient's health status can be obtained from this data, enabling the early identification of anomalies or changes that might point to possible health issues. Wearable technology that tracks pertinent health indicators might be helpful for patients with long-term diseases like diabetes, hypertension, or heart problems. These devices offer actionable data for therapy adjustments and lifestyle alterations, which can help patients and healthcare practitioners manage their ailments more successfully. Wearable technology makes it possible to remotely monitor patients outside of conventional hospital settings, which enables more proactive and individualized care. Wearable devices can also identify falls in elderly or at-risk people and immediately notify emergency services or caregivers using sensors [18–22]. In the event of a fall-related injury, this capability might greatly speed up reaction times, possibly saving lives and lessening the severity of injuries.

Wearable fitness trackers are becoming more and more common for keeping tabs on exercise performance, physical activity levels, and general wellness. Better health outcomes and the avoidance of disease can be encouraged by using these devices to track progress over time, create and accomplish fitness goals, and maintain an active lifestyle. Smart wearable devices can also monitor and remind patients to follow their recommended medication schedules. These devices can increase treatment efficacy and lower the risk of consequences related to non-compliance by promoting medication adherence. Patients and healthcare professionals can have remote consultations due to the integration of wearable technology with telemedicine systems. Without requiring in-person appointments, medical practitioners can evaluate wearable data in real time, talk about treatment options, and offer individualized health recommendations during these virtual visits.

Mechatronics has had a considerable impact on medical imaging systems. The development of the precision and accuracy of imaging devices is one of the most significant contributions that mechatronics has made to the field of medical imaging [23]. By incorporating sensors, actuators, and sophisticated control systems, mechatronics-based medical imaging systems are able to achieve extraordinary performance. Furthermore, the integration of robots and computer systems enables advanced image analysis and processing, such as three-dimensional reconstructions and real-time image-guided interventions, which expands the possibilities of medical imaging beyond the limits of conventional approaches. Mechatronics has been a significant contributor to the development of medical imaging systems that are more dependable, efficient, and capable of giving full data for the purposes of medical analysis and operations. MRI (Magnetic Resonance Imaging), CT (Computed Tomography), ultrasound, and PET (Positron Emission Tomography) scanners are examples of sophisticated medical imaging systems that have been developed due to mechatronics. Non-invasive viewing of internal organs and tissues is made possible by these systems, which combine mechanical elements for patient positioning, electronic sensors for data collecting, and computer algorithms for image processing.

Mechatronic systems are crucial in robotically assisted surgery because of the delicate nature of the instruments used and the importance of minimizing patient trauma. Surgical robots are the result of interdisciplinary efforts in mechanical, electrical, computer, and control engineering [24]. Mechatronics systems also enable the development of exoskeletons and enhanced prosthetic limbs for individuals afflicted with mobility impairments or limb loss. Devices with built-in sensors, actuators, and control systems can restore mobility, improve users' quality of life, replicate natural movements, and improve their abilities [25]. Last but not the least, when it comes to pharmaceutical interventions, mechatronic drug delivery devices are a significant addition [26]. Improved efficacy and safety of drug administration are now possible by the precise control of these systems overdosage and release mechanisms.

4.4 Mechatronics in Agriculture

Recently, mechatronics in agricultural sector has been the subject of much research [27, 28]. Mechatronics has revolutionized farming by improving agricultural operations integrating electrical, mechanical, and computer engineering. Mechatronics has been a huge boon to precision farming, autonomous farming, robotic systems, and crop monitoring and spraying drones. By automating planting and harvesting systems, precision irrigation, and nutrient delivery, mechatronics has improved agricultural efficiency and productivity while lowering resource requirements and environmental impact, leading to higher crop yields and reduced labor costs.

4.5 Mechatronics in Transportation

The advancement of safer, more efficient, and environmentally friendly transportation systems and vehicles is made possible by the technical revolution in mechatronics, which is crucial to the transportation industry. The multidisciplinary approach enables engineers to address the intricate obstacles that stand in the way of intelligent, autonomous mobility as a whole. Mechatronics has a profound impact, especially in the domains of intelligent transportation systems, self-driving cars, and self-driving technology [29].

4.6 Mechatronics in Aerospace

Mechatronics has numerous applications in the development of unmanned aerial vehicles (UAVs) and drones, as well as in the deployment and maintenance of satellites, automation in aerospace production, and space exploration in the aerospace industry. The integration of numerous technologies has spurred enormous developments in the field of aerospace and space exploration over the years [30]. A key component of this advancement has been mechatronics, which enables the development of autonomous and intelligent systems capable of operating in harsh conditions and carrying out intricate operations in the aerospace and space exploration industries with remarkable efficiency and accuracy.

4.7 Mechatronics in Energy

The application of mechatronics is immensely essential in the field of smart grid technology that distributes energy in an efficient manner as well as in renewable energy systems such as solar, wind, and hydropower. Smart grid technology distributes energy in an efficient manner. The technological advancements in the field of mechatronics play a significant role in the development and implementation of smart grid technology as well as renewable energy systems [31].

5 Advancements in Mechatronics: Harnessing the Power of AI for Intelligent Systems

Artificial Intelligence (AI) is a remarkable approach toward the development and implementation of Mechatronics. Control technology plays a major role in the design and realization of this type of intelligence; that is, mechatronic systems especially actualize control systems. To keep a process or its surroundings in the best possible condition, the system should be able to pick up a variety of innate facts and information. After that, it ought to apply what it has learned to estimation, analysis, control, and decision-making. It is capable of providing the proper care when it comes across anything it hasn't learned. When there is a local failure, the system can still function, even analyzing and fixing the issue. The system possesses a great deal of adaptability and initiative. It can act independently within the parameters of the task; in the event of a disagreement, the controller can be instructed to resolve it, among other things. These days, the primary fields of study in artificial intelligence are neural networks, fuzzy logic, expert systems, neural network control, learning control, and hierarchical progression. Such fields are developing independently until they finally combine. Electromechanical systems, which are now utilized to achieve intelligence, are most advanced in the first three of these. When an input signal is received, an artificial neural network can mimic the high degree of connectivity found in many human brain cells. This allows neurons to be activated and then processed through a neural circuit to produce an output. Because of its associative memory and learning capabilities, a neural network can generate the desired output following an input signal. An information circuit can also generate a respectable output if it hasn't been taught.

Applications of artificial neural networks in mechatronic systems have advanced significantly, and combining them with fuzzy logic and expert systems is an important area for future research. Fieldbus LON WORKS is a chip designed for mechatronic systems that use neuron chips as their primary technology. It can finish the ISO/OSI reference model of the 3–6 layer network protocol, user field control applications, and media access control (MAC) function with its three internal microprocessors. This completes the application for the network processor. They exchange data by means of a shared memory. Several V0 and time calculators are also included in the neuron chip's features. Data and control are combined on a tiny neuron chip, which also has strong communication capabilities. In a dispersed control system, this chip and a few other devices may sometimes function as independent control units. In summary, each of the three technologies—expert systems, fuzzy logic, and artificial neural networks—has a specific role to play in the ongoing, well-coordinated development of new technologies that elevate the bar for mechatronic system intelligence and expand its potential applications.

6 Impacts of Microelectronics Technology

Many facets of daily life are still expanding as a result of the rapid rise of civilization in the modern era. Through the use of digital management systems and contemporary microelectronics technology, mechatronics has become such a system that it dominates the field of electromechanics. The rapid advancement of contemporary technology like microelectronic controllers and computer applications forms the foundation of the new digital system of the mechatronics era. The application of contemporary computer technologies has also made integrated management and virtualization conceivable. Mechatronics technology is an essential tool in the growth of the mechanical and electrical industries since it allows these products to have digital characteristics. In the current era, electronic information technology has made substantial advancements in mechatronics digital management systems, increasing its capacity to improve the digital production of mechatronics digital management systems. With the help of many programmable controllers and digital management systems, mechatronics can advance intelligently in the manufacturing of mechanical and electrical items. Mechatronics has evolved into intelligence in the modern period, and as time passes, this development is progressively influencing our nation.

Working together, professionals from several design areas are necessary for the development of mechatronic systems. However, a number of difficulties associated with the design and development of a mechatronic system have been addressed by researchers [32]. The automation of present design procedures and the integration of design and analytic technologies are the main areas of difficulty. By addressing these issues, a concurrent development strategy that considers the synergistic effects characteristic of mechatronic systems in the design process can be adopted. The main argument is that other than concurrency, system design techniques need to be considered in order to overcome problems associated with software development for complex mechatronic systems. It is necessary to consider the complex inter dependencies that exist between subsystems and the designers that construct them. To find potential answers put out in earlier publications, an evaluation of the available techniques and resources is conducted. The goal of the research is not to do a thorough analysis, but rather to demonstrate that integration is a significant problem from several angles and that overcoming integration obstacles can be accomplished by raising the abstraction level in system descriptions. There is also a brief analysis that highlights the challenges in developing mechatronic products, which are predicated on a growing degree of abstraction. Depending on such outcomes, ideas for an integration framework are put forth. The framework's objective is to assist a multidisciplinary design team in producing and validating control software almost entirely autonomously. Knowledge-based techniques and technologies can assist in the software development and verification process based on high-level architectural descriptions. Additional objectives include fostering engineer-to-engineer collaboration, enhancing design reliability, maximizing the reuse of design information, and cutting down on development expenses and time [33].

7 Technology for Better Standard of Living

Industrial technology and human civilization have produced intelligent machines, high-tech tools, equipment, and goods that raise living standards [34]. Using mechatronics, intelligent machines with conscience and decision-making abilities can be built. In order to generate new and technologically enabled goods that meet client expectations, companies increasingly require high-tech feedback. The perfect integration of mechanics and technology enables the design and production of futuristic vehicles and gadgets like smartphones. Mechatronics have improved performance and product quality in advanced manufacturing. By using automated systems, simplified processes can be used more smoothly and time can be saved as well. Industrial production has become more affordable and productive due to mass manufacturing [35]. By using intelligent sensors [36–39] and parameter controls to preserve the intended operating conditions [39–41], technical faults are reduced to a minimum. This technology goes far beyond creating electromechanical systems that are controlled by computers. Over time, mechatronics has expanded to encompass many more fields, such as telecommunication [42–44], computer technology [45–47], systems engineering [48, 49], Internet of Things [50–52], civil engineering [53], and instrumentation and control [54]. Originally, mechatronics was thought to be an amalgamation of mechanics and electronics. The complexity of systems and devices has skyrocketed in recent years, necessitating a system-level approach to mechatronics design. This method facilitates the easy execution of tests, the blending of mechanical and control design, and the reuse of algorithms inside the embedded delivery framework. Systemic advancements in industrial technology and growing investments in the biosciences, medicine, and renewable energy sectors are driving this development.

The design process is significantly improved when widely available technology is combined with industry best practices to speed up design, prototyping, and implementation. Through the division of the design process into multiple concurrent processes, the engineers might perhaps devise a more efficient approach. A team had to wait for a functional prototype before developing a control algorithm for a mechanical device. Using a virtual prototype made from concept models and simulation data, engineers can now work more swiftly. This method reduces development time and related expenses while fostering active collaboration among design teams and offering early-stage input on system performance. Teams can increase performance and mutually automate the overall structure with a full system digital model. However, a well-designed mechatronics design process comes with integrated design tools. Device vendors have made significant investments in promoting this design approach in the last few years. Companies such as National Instruments have developed interfaces to enable smooth communication between different devices. Because of this, engineers may more easily develop virtual versions of their systems that can be easily transferred to embedded hardware, reuse code, and get the most out of their investment in the design and simulation stages. When developing and implementing algorithms, embedded hardware must be used. During deployment, engineers can reduce production time in a number of ways. This include developing an embedded commercial-off-the-shelf (COTS) control system that is modified at the component level to reduce costs [55]. Making the most logical choice based on these trade offs is what separates a successful investor from a loser. "System ready" subsystems are mostly employed in delivery platforms or prototypes to gauge user acceptability prior to designing a specific design that maximizes cost.

A trend for configurable embedded platforms has emerged in recent years, providing both the rapid deployment time of COTS hardware and the flexibility of a custom configuration. About seven years ago, National Instruments released RIO (reconfigurable I/O). Customers might incorporate software algorithms into hardware with a programmable plug-in board for PCs that has an FPGA and integrated analog and digital I/O by using the graphical programming environment NI LabVIEW. Durable embedded hardware and industrial-scale system-level devices currently use this technology, which has advanced since its inception [56]. Some researchers focused on the automation of everything in the context of the Industrial Revolution and proposed that this represents a shift in complex mechatronic systems [57]. Future embedded systems and mechatronics will heavily rely on software. Software tools assist engineers in creating high-performance systems capable of meeting future difficulties by providing a system-level perspective of several design characteristics and abstracting the subtleties of cutting-edge technology.

8 Conclusion

The advancement of human civilization through technology along with the emerging trends in mechatronics will raise awareness of what lies ahead. A more robust system will result from the highlighted technologies' incorporation and integration. Mechatronics has already paved the way for a number of intriguing future directions. Indeed, the adoption of a traditional single discipline method or a classic sub-system would never have allowed mechatronics to open up such significant technical options. Technological developments impact every industry and have an impact on everyone. Practitioners of data science will be successful regardless of their prior professional experience. Furthermore, mechatronic technologies are improving the standard of living in society and revolutionizing fields like medicine. Due to the skills of data scientists, advancements in machine learning (ML) and artificial intelligence (AI) are currently changing the marketplace. Effective mechatronic design will produce reasonably priced goods that are very appealing to customers. Products created in a more conventional and sequential manner seem outdated, have few features, and are not very appealing to consumers. Numerous diverse fields of knowledge are supported by mechatronics. In order to keep up with the constantly shifting demands and requirements of a dynamic and complicated society, inventions and technologies must advance quickly. Mechatronics will be a key component in improving product affordability, protection, and dependability in the near future. Further investigations can examine the adverse impacts of these technologies and the essential remedies to alleviate the risks.

References

1. Singh AK, Saxena A, Gupta M (2017) Study and analysis of use of mechatronics in modern machine design. Int J Emerg Technol (Special Issue NCETST-2017) 8(1):345–354
2. Li Y, Huang J, Tang H (2012) A compliant parallel XY micro-motion stage with complete kinematic decoupling. IEEE Trans Autom Sci Eng 9(3):538–553
3. Yun Y, Li Y (2011) Optimal design of a 3-PUPU parallel robot with compliant hinges for micromanipulation in a cubic workspace. Robot Comput-Integr Manuf 27(6):977–985
4. Tang Y, Li Y (2011) The software architecture of a reconfigurable real-time onboard control system for a small UAV helicopter. In: Proceedings of the 8th international conference on ubiquitous robots & ambient intelligence (URAI '11). Songdo Convention, Incheon, Korea, pp 228–233
5. Tang Y, Li Y (2011) Development of a laboratory HILs testbed system for small UAV helicopters. In: Proceedings of the IASTED international conference on robotics (Robo '11). Pittsburgh, Pa, USA, pp 428–436
6. Wang J-G, Li Y (2010) Hybrid impedance control of a 3-DOF robotic arm used for rehabilitation treatment. In: Proceedings of the 6th IEEE international conference on automation science and engineering (CASE '10). Ontario, Canada, pp 768–773
7. Shetty D, Manzione L, Ali A (2012) Survey of mechatronic techniques in modern machine design. J Robot 2012:9
8. Xiang Y (2021) Exploration of the application of artificial intelligence technology in mechatronics technology based on. In: Journal of physics: conference series, ISCE 2021, vol 1915, p 022059
9. Zhen LY (n.d.) Multimedia assisted English teaching inquiry country. J Human Inst Sci Technol 206(8)
10. (2023) Innovation trends in additive manufacturing. Patents in 3D printing technologies
11. Pose-Rodriguez J, Ménière Y, Philpott J, Rudyk I, Wewege S, Wienold N (2020) Patents and the fourth industrial revolution: the global technology trends enabling the data-driven economy
12. Saxena M, Jain P, Tanvar A, Singh P, Bansal N, Kumar A (2020) Emerging trends in mechatronics with AI: a review. J Crit Rev 7(5)
13. Khan WM (2023) How mechatronics is essential in future for living a flexible and reliable life. 4(11)
14. Arshad A, Khan S, Midi NS, Ismail F, Tasnim R, Islamic I (2016) Daily activity monitoring of an elderly Person for determining their wellness. Int J Smart Home 10:339–346
15. Arshad A, Khan S, Alam AZ, Ahmad FI, Tasnim R (2014) A study on health monitoring system: recent advancements. IIUM Eng J 15(2)
16. Han SA, Naqi M, Kim S, Kim JH (2022) All-day wearable health monitoring system. EcoMat 4(4):e12198
17. Arshad A, Ismail AF, Khan S, Alam AZ, Tasnim R, Haider SS, Shahid Z (2013) A qualitative review for wireless health monitoring system. In: IOP conference series: materials science and engineering, vol 53, No. 1. IOP Publishing, p 012030

18. Arshad A, Khan S, Alam AZ, Ismail AF, Tasnim R (2017) Capacitive proximity floor sensing system for elderly tracking and fall detection. In: 2017 IEEE 4th international conference on smart instrumentation, measurement and application (ICSIMA), pp 1–5
19. Singh A, Rehman SU, Yongchareon S, Chong PHJ (2020) Sensor technologies for fall detection systems: a review. IEEE Sens J 20(13):6889–6919
20. Arshad A, Tasnim R, Alam AZ, Khan S (2015) Capacitance-to-voltage converter design to measure small change in capacitance produced by human body movement. In: 2015 IEEE international WIE conference on electrical and computer engineering (WIECON-ECE), pp 114–117
21. Taramasco C, Rodenas T, Martinez F, Fuentes P, Munoz R, Olivares R, Demongeot J (2018) A novel monitoring system for fall detection in older people. Ieee Access 6:43563–43574
22. Arshad A, Tasnim R, Alam AMZ, Khan S (2016) Low value capacitance measurement system for the application of monitoring the human body. ARPN J Eng Appl Sci 11(1):327–330
23. Cynober M, Saito Y, Chaillet N (2012) Development of an automatic and mobile X-ray robot: an innovative approach to medical imaging through mechatronics. J Franklin Inst 349(7):2313–2322
24. Alterovitz R, Desai JP (2009) Surgical robotics [TC spotlight]. IEEE Robot Autom Mag 16(2):16–17
25. Laschowski B, McPhee J, Andrysek J (2019) Lower-limb prostheses and exoskeletons with energy regeneration: mechatronic design and optimization review. J Mech Robot 11(4):040801
26. Munoz F, Alici G, Li W (2016) A magnetically actuated drug delivery system for robotic endoscopic capsules. J Med Dev 10(1):011004.10
27. Azeta J, Bolu CA, Alele F, Daranijo EO, Onyeubani P, Abioye AA (2019) Application of mechatronics in agriculture: a review. In: Journal of physics: conference series, vol 1378, No. 3. IOP Publishing, p 032006
28. Najeh LM, Ala BALTI (2021) The need for system approach and mechatronics for agricultural applications. J Comput Sci Control Syst 14(1):26–29
29. Wójcik W, Pavlov S, Kalimoldayev M (eds) (2021) Mechatronic systems 1: applications in transport, logistics, diagnostics, and control. Routledge
30. da Fonseca IM, Pontuschka MN (2015) The state-of-the-art in space robotics. In: Journal of physics: conference series, vol 641, No. 1. IOP Publishing, p 012025
31. Dogra V (2021) Design and optimization of mechatronic systems for renewable energy harvesting. Math Stat Eng Appl 70(1):371–377
32. Cabrera AAA, Foeken MJ, Tekin OA, Woestenenk K, Erden MS, De Schutter B, van Tooren MJL, Babuska R, Houten FJAM, Tomiyama T (2010) Towards automation of control software: a review of challenges in mechatronic design
33. Schöner HP (2004) Automotive mechatronics. Control Eng Pract 12(11):1343–1351
34. Nnodim TC, Arowolo MO, Agboola BD, Ogundokun RO, Abiodun MK (2021) Future trends in mechatronics. IAES Int J Robot Autom 1(10):24
35. Jaradat O, Sljivo I, Habli I, Hawkins R (2017) Challenges of safety assurance for industry 4.0. In: 2017 13th European dependable computing conference (EDCC), Geneva, pp 103–106. https://doi.org/10.1109/EDCC.2017.21
36. Nnodim CT, Fath El-Bab AMR, Ikua BW, Sila DN (2019) Design and simulation of a tactile sensor for fruit ripeness detection. In: Lecture notes in engineering and computer science: world congress on engineering and computer science 2019, pp 390–395
37. Detweiler C, Doniec M, Vasilescu I, Rus D (2012) Autonomous depth adjustment for underwater sensor networks: design and applications. IEEE/ASME Trans Mechatron 17(1):16–24. https://doi.org/10.1109/TMECH.2011.2175003
38. Suryadevara NK, Mukhopadhyay SC, Kelly SDT, Gill SPS (2015) WSN-based smart sensors and actuator for power management in intelligent buildings. IEEE/ASME Trans Mechatron 20(2):564–571. https://doi.org/10.1109/TMECH.2014.2301716
39. Hur H, Ahn H (2014) Unknown input observer-based filterings for mobile pedestrian localization using wireless sensor networks. IEEE Sens J 14(8):2590–2600. https://doi.org/10.1109/JSEN.2014.2312193

40. Zhou X, Beilei Z, Guohao G (2015) Control parameters optimization based on co-simulation of a mechatronic system for an UA-based two-axis inertially stabilized platform. Sensors 15(8):20169–20192
41. Ma J, Chen SL, Teo CS, Tay A, Al Mamun A, Tan KK (2019) Parameter space optimization towards integrated mechatronic design for uncertain systems with generalized feedback constraints. Automatica 105:149–158
42. Parvez I, Rahmati A, Guvenc I, Sarwat AI, Dai H (2018) A survey on low latency towards 5G: RAN, core network and caching solutions. IEEE Commun Surv Tutor 20(4):3098–3130. https://doi.org/10.1109/COMST.2018.2841349
43. Boccardi F, Heath RW, Lozano A, Marzetta TL, Popovski P (2014) Five disruptive technology directions for 5G. IEEE Commun Mag 52(2):74–80. https://doi.org/10.1109/MCOM.2014.6736746
44. Gupta A, Jha RK (2015) A survey of 5G network: architecture and emerging technologies. IEEE Access 3:1206–1232. https://doi.org/10.1109/ACCESS.2015.2461602
45. Schraft RD (1994) Mechatronics and robotics for service applications. IEEE Robot Autom Mag 1(4):31–35
46. Komoto H, Tetsuo T (2012) A framework for computer-aided conceptual design and its application tosystem architecting of mechatronics products. Comput Aided Des 44(10):931–946
47. Ai Y, Peng M, Zhang K (2018) Edge computing technologies for internet of things: a primer. Digital Commun Netw 4(2):77–86
48. Madni AM, Madni CC, Lucero SD (2019) Leveraging digital twin technology in model-based systems engineering. Systems 7(1):1–13
49. Matar M, Osman H, Georgy M, Abou-Zeid A, El-Said M (2017) A systems engineering approach for realizing sustainability in infrastructure projects. HBRC J 13(2):190–201
50. Sheng Z, Mahapatra C, Zhu C, Leung VCM (2015) Recent advances in industrial wireless sensor networks toward efficient management in IoT. IEEE Access 3:622–637
51. Chiang M, Zhang T (2016) Fog and IoT: an overview of research opportunities. IEEE Internet Things J 3(6):854–864
52. Delsing J (2017) Local cloud internet of things automation: technology and business model features of distributed internet of things automation solutions. IEEE Ind Electron Mag 11(4):8–21
53. Gattulli V, Ottaviano E, Pelliccio A (2018) Mechatronics in the process of cultural heritage and civil infrastructure management. In: Intelligent systems, control and automation: science and engineering, vol 92. Springer International publishing AG, pp 1–31
54. Santhosh TV, Gopika V, Ghosh AK, Fernandes BG (2018) An approach for reliability prediction of instrumentation & control cables by artificial neural networks and Weibull theory for probabilistic safety assessment of NPPs. Reliab Eng Syst Saf 170:31–44
55. St. Onge P, Temme LA, McAtee LA, O'Brien KJ, Byrd BK (2019) Evaluation of the commercial, offthe-shelf (COTS) King-Devick eye tracking system. Defence Tech Inform Center 184(1):571–578. https://doi.org/10.1093/milmed/usy380
56. Hajduk Z (2017) High accuracy FPGA activation function implementation for neural networks. Neurocomputing 247:59–61
57. Kuru K, Yetgin H (2019) Transformation to advanced mechatronics systems within new industrial revolution: a novel framework in automation of everything (AoE). IEEE Access 7:41395–41415

Ethical and Social Consideration in Mechatronics

Md. Hosne Mobarak, Md. Jahid Hasan, Mustafizur Rahman, Masud Rana, Abu Salman Saikat, and Rumana Tasnim

Abstract This chapter explores the ethical and social aspects of mechatronics, with a focus on the responsible creation and utilization of technology. From an academic perspective, it emphasizes the importance of innovations conforming to ethical principles, with a focus on ensuring safety, privacy, and sustainability. It emphasizes the importance of mechatronics in meeting society's demands, reducing disparities, and improving quality of life by promoting transparent and inclusive decision-making from a social perspective. Ethical frameworks such as utilitarianism, deontology, virtue ethics, and rights-based ethics serve as a basis for developing equitable and socially responsible technologies. The chapter also examines the societal consequences of mechatronics, including economic upheavals and the possibility of job displacement, while acknowledging its ability to stimulate innovation and generate fresh prospects. Case studies in robotics, such as autonomous automobiles and healthcare robots, demonstrate the ethical intricacies and societal consequences of these technologies. The chapter promotes the importance of ethical concerns in the development of mechatronic systems and cyber-physical-Human systems (CPHS), with a focus on human-centered design. The text also explores the significance of regulatory frameworks, international standards, and educational programs in promoting responsible innovation. These measures ensure that technical progress is in line with society's values and makes a constructive contribution to global development.

Md. H. Mobarak · M. Rahman · M. Rana (✉) · R. Tasnim
Department of Mechatronics Engineering, World University of Bangladesh, Dhaka, Bangladesh
e-mail: masud.ranamte@mte.wub.edu.bd

Md. J. Hasan
Department of Mechanical Engineering, DUET, Gazipur City, Bangladesh

A. S. Saikat
Bharti School of Engineering and Computer Science, Laurentian University, Sudbury, Canada

M. M. Rahman et al. (eds.), *Mechatronics*, Emerging Trends in Mechatronics,
https://doi.org/10.1007/978-981-97-7117-2_8

1 Introduction

In mechatronics, ethical and social considerations include the conscientious creation and utilization of technology. From an academic perspective, it entails ensuring that innovations conform to ethical principles, giving priority to safety, privacy, and sustainability. From a social perspective, mechatronics should focus on meeting society's needs, minimizing disparities, and improving the overall quality of life. This necessitates the implementation of transparent and inclusive decision-making procedures, as well as the ongoing evaluation of the effects of technology on communities. Ensuring a harmonious alignment between technological progress and moral responsibilities cultivates confidence and facilitates the advantageous assimilation of mechatronics into our everyday existence.

In mechatronics, ethical frameworks provide guidance for the proper creation and use of technology [1]. Utilitarianism prioritizes the maximization of general satisfaction and the minimization of harm, with the aim of ensuring that mechatronic advancements are beneficial to the majority [2, 3]. Deontological ethics prioritizes obligation and strict devotion to moral principles, advocating for the unwavering application of ethical laws regardless of the resulting consequences [4]. Virtue ethics places emphasis on the cultivation of moral character, urging engineers to exemplify characteristics such as honesty, integrity, and accountability. Rights-based ethics emphasizes the protection of individual rights, ensuring that advancements in mechatronics adhere to and support human dignity and freedom [5, 6]. By incorporating these ethical viewpoints, it is possible to develop technologies that are both fair and socially accountable, thereby fostering confidence and generating beneficial effects on society [7].

Mechatronics has significant social ramifications that affect multiple facets of society. The advent of automation and modern technology is causing economic disruption by fundamentally transforming industries, which in turn may lead to the displacement of labor and the need to acquire new skill sets [1, 8]. This can worsen social inequality if there is uneven access to education and retraining [9]. On the other hand, mechatronics has the potential to generate employment prospects in advanced industries and enhance technology accessibility, hence promoting innovation [10]. Societal adjustments to new technological standards give rise to cultural and behavioral transformations [11]. Furthermore, it is imperative to take into account the environmental repercussions of mechatronics, as the implementation of sustainable measures is crucial in order to alleviate any adverse consequences [12]. Ensuring a harmonious equilibrium of these aspects is crucial in order to guarantee that mechatronics makes a good contribution to society, fostering equitable growth and development [13].

Case studies in robotics focus on the ethical concerns surrounding the social implications of autonomous vehicles and the ethical challenges faced by healthcare robotics. Autonomous vehicles give rise to concerns regarding safety, dependability, and responsibility since they can prevent fatalities but also carry the possibility of crashes and malfunctions [14–16]. Healthcare robotics must strike a balance between advancing technology and upholding ethical principles related to patient care and privacy. Industrial automation raises ethical problems around job displacement and its economic consequences, requiring the development of methods to facilitate workforce adaptability [17, 18]. Ensuring that robot technologies benefit society and effectively solve ethical challenges is of utmost importance, and human-centered design plays a critical role in achieving this goal [19]. These case examples emphasize the significance of adopting a comprehensive approach to robotics, taking into account both technological progress and its wider societal consequences [20].

Developing mechatronic systems and cyber-physical-Human systems (CPHS) necessitates the incorporation of ethical design and responsible innovation [21]. These technologies can benefit society and mitigate potential negative consequences by incorporating ethical considerations [22]. Case studies highlight the significance of integrating ethics into engineering education, giving upcoming engineers the ability to tackle ethical challenges [23]. Responsible research and innovation techniques are crucial for fostering sustainability and enhancing societal well-being [24]. By prioritizing these elements, we can create systems that are not just technologically sophisticated but also morally upright, demonstrating a dedication to conscientious progress and its wider effects on society [25].

The regulatory and policy aspects of mechatronics involve the implementation of international standards and guidelines to guarantee uniformity and safety across different countries [26]. Government rules and oversight are essential for ensuring compliance and safeguarding the public interest [27]. The implementation of ethical standards of conduct is crucial for professionals in the field of mechatronics, as they serve to foster integrity and accountability [28]. These frameworks provide experts with guidance on following established standards, ensuring that technological progress benefits society while limiting potential risks [29]. By incorporating these components, the mechatronics sector can responsibly promote innovation, conforming to both domestic and global standards. Academic institutions also have a crucial role in imparting these ideas to future professionals [30, 31].

2 Ethical Frameworks in Mechatronics

An interdisciplinary branch of engineering called mechatronics focuses on combining software, electrical, mechanical, and electronic engineering. This chapter proposes a conceptual framework for considering ethical issues in mechatronic engineering. Ethical frameworks offer standards and principles to help ensure that professionals in the field make decisions that prioritize ethical concerns. Mechatronics is a field that spans robotics, computer science, telecommunications, systems, control, and product engineering. Because of its interdisciplinary character, mechatronics presents particular ethical concerns. Some key ethical frameworks include utilitarianism, deontology, virtue ethics, rights-based ethics, sustainable development, and engineering codes of ethics that are relevant to mechatronics.

2.1 Utilitarianism

A consequentialist ethical theory known as utilitarianism determines the morality of an act by considering its effects or possibilities. Regardless of its best efforts, utilitarian ethics is unable to overcome a number of fundamental defects. Utilitarian ethics maintains the moral justification of a behavior or policy, even when it violates the rights of others. Utilitarian ethics completely justifies the violation of someone's rights if it results in something positive. Gains and losses cannot be quantified. It is impossible to tell whether a given action or policy creates a game with a positive total. It is only possible to determine who won and who lost [32]. Lottu et al. [33] suggested the inclusion of ethical issues at every stage of the AI development lifecycle, including design, implementation, deployment, and maintenance. Utilitarianism places a high value on maximizing the overall utility or advantage to society, which can serve as a useful framework for making ethical decisions. Philosophers like Jeremy Bentham and John Stuart Mill introduced the consequentialist theory of utilitarianism. It argues that the optimal course of action is the one that maximizes overall happiness or utility. Within the realm of mechatronics, engineers can rely on utilitarian concepts to make judgments that maximize the overall benefit for the largest possible number of individuals [34].

Mechatronics engineers can apply utilitarianism by:

- Assessing the potential advantages and disadvantages of their projects and selecting those that have the greatest overall good effect.
- Focusing on projects and research with the potential to greatly enhance quality of life, such as medical robots or sustainable automation solutions.
- Creating products that improve safety, accessibility, and overall well-being for a wide range of people.
- Creating robotic systems that enhance surgical procedures and patient care, ultimately leading to improved healthcare outcomes for a significant patient population.
- Developing automated solutions for pollution control or renewable energy, thereby promoting environmental sustainability and yielding societal benefits.

2.2 Deontological Ethics

Deontological ethics determines the morality of an action based on its adherence to rules or principles, not its consequences [35]. This section assesses how well moral guidance encourages people to make moral decisions using the deontological ethics framework. Kim et al. [36] looked at how moral counsel based on deontological ethics frameworks may effectively assist humans in making moral decisions and assessed how well robots can perform this role for people. The focus of deontological ethics is on moral action that conforms to established, broadly applicable moral precepts [37]. Mechatronics can apply deontological ethics in a variety of ways, such as adhering to safety standards, respecting human autonomy, maintaining privacy and confidentiality, maintaining integrity in design, upholding property rights, designing for robustness and reliability, and using technology ethically. Deontological ethics, based on Immanuel Kant's philosophy [38], emphasizes the strict observance of duty, norms, and responsibilities without considering the outcomes. This ethical framework places a significant emphasis on adhering to ethical principles and professional standards in the field of mechatronics [3].

Deontological ethics should guide mechatronics engineers in the following ways:

- Adhere to recognized engineering codes of ethics, such as IEEE and ASME, which prescribe ethical and professional conduct.
- Prioritize safety, honesty, and integrity in all engineering practices, regardless of any external demands or potential rewards.
- Acknowledge and uphold the independence of users and clients, guaranteeing their informed agreement and clear communication.
- Follow safety regulations and processes during the design and testing of new robotic systems, even if taking shortcuts could speed up development.
- Ensure that precise and honest reports and assessments are provided, avoiding any deception or manipulation of data.

2.3 *Virtue Ethics*

Applying virtue ethics' tenets to the conception, creation, and use of mechatronic systems [39]. Virtue ethics values individual character, emphasizing qualities like accountability, honesty, and integrity. It fosters responsible innovation, ensuring the creation and application of mechatronic technologies in a manner that upholds moral principles and benefits society [40]. Additionally, it pushes engineers to give human dignity and well-being a top priority during the design phase. In the creation and use of mechatronic systems, virtue ethics in mechatronics highlights the value of morality, ethical design, responsible innovation, human-centeredness, professional integrity, and ongoing improvement. Aristotle created the philosophical paradigm of virtue ethics, emphasizing the importance of cultivating virtuous character attributes. This entails fostering virtues such as integrity, bravery, equity, and accountability among engineers in the field of mechatronics [41].

Promoting virtue ethics in mechatronics involves the following:

- Promoting education and training initiatives that prioritize the cultivation of both technical expertise and ethical values.
- Offering mentorship opportunities where seasoned engineers exemplify virtuous conduct and ethical decision-making.
- Entails creating a work environment that values and acknowledges virtuous actions, encouraging consistent ethical behavior over an extended period of time.
- Pursuing innovation in a manner that aligns with ethical principles, such as creating technologies that do not take advantage of users or communities,

- Demonstrating the bravery to voice opposition to unethical behavior within organizations, especially while facing potential personal or professional consequences.

2.4 Rights-Based Ethics

In mechatronics, rights-based ethics refers to the application of moral precepts that give individual liberties and rights top priority during the design, development, and operation of mechatronic systems [42]. It is the responsibility of engineers to guarantee that mechatronic systems adhere to and preserve human rights, such as freedom of speech, privacy, and security against damage. This entails creating procedures that respect these rights without violating them [43]. Professionals in the sector should make judgments that prioritize ethical considerations by following the standards and principles provided by ethical frameworks [44]. However, the efficacy of moral communication based on these theories may differ. It may be varyingly effective in different circumstances and with different people because these theories emphasize different methods of moral reasoning [45]. Rights-based ethics, rooted in the philosophy of prominent thinkers such as John Locke and further refined in contemporary human rights theory, centers on safeguarding and honoring the rights of individuals [46]. This ethical framework places significant emphasis on protecting the rights of all individuals and groups involved in the field of mechatronics [47].

Mechatronics engineers should consider the following:

- Ensure that products and systems uphold users' privacy rights and safeguard sensitive data.
- Develop technologies that ensure inclusivity and equitable opportunities for people with diverse abilities.
- Before using or implementing mechatronic systems, mechatronics engineers should thoroughly brief users and stakeholders on their functionality and potential hazards.
- Develop robotic systems, particularly those used in personal environments that meticulously safeguard user privacy and adhere to data protection rules.
- Develop inclusive products that are accessible to a diverse range of individuals, including those with impairments, in order to uphold their right to technology.

3 Social Implications of Mechatronics

Mechatronic devices are an often-used category of items. Applications for mechatronic devices range widely, from standard consumer devices like DVD players to intricate and crucial subsystems found in airplanes. Given the prevalence of mechatronic devices, it is necessary to design them cyclically. Hence, the objective of this study is to determine the present status of circular design in mechatronic development, with the aim of enhancing comprehension and implementation of a circular

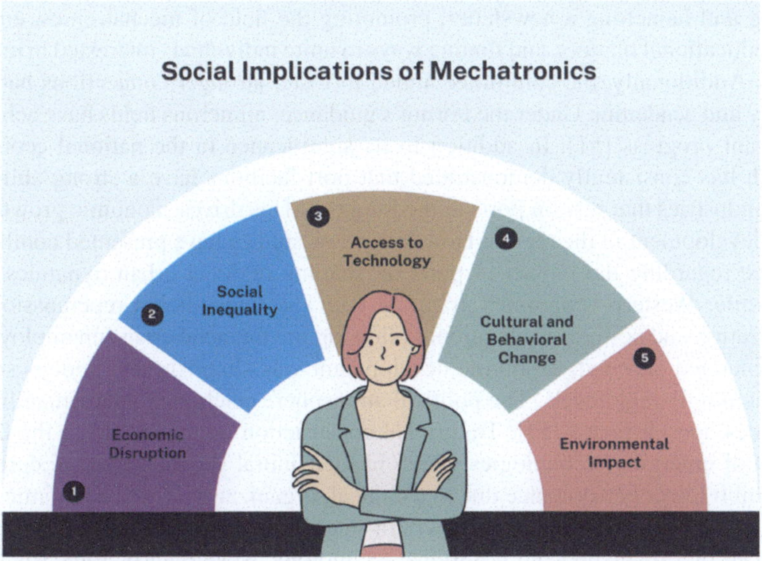

Fig. 1 Social implications of mechatronics

economy in mechatronics design [48]. As a result, the industrial world is currently undergoing a rapid transition from the previous industrial revolution, known as Manufacturing 3.0, which was characterized by electronics, computing, and automation. This transition is leading toward another industrial revolution known as Industry 4.0, which is characterized by digitalization, cloud computing, the Internet of Things, and cyber-physical systems (CPS). Additionally, there is a slower transition phase, which involves the displacement of humans from production systems [49]. Figure 1 provides a discussion on the societal ramifications of mechatronics. The following are the six primary topics we will discuss: This explores economic disruption, social inequality, technology access, and the impact of cultural and behavioral changes on the environment.

3.1 *Economic Disruption*

An industrial revolution is defined as a period of significant technological advancements that improve the process of manufacturing. The first industrial revolution began by utilizing water and steam power in the production process. The adoption of assembly line techniques and the use of electric power in manufacturing marked the achievement of the second stage of industrialization [50]. The Mechatronics Forum gave its initial committee a wide range of responsibilities, including creating and launching a newsletter, promoting the field of mechatronics, emphasizing educational matters, and finding ways to unite individuals interested in mechatronics. Additionally, the committee aimed to foster stronger connections between industry and academia. Under the Forum's guidance, numerous fields have achieved significant progress [51]. In addition to its significance in the national economy, research has consistently demonstrated that port facilities have a strong ability to attract industries that rely on ports in the long run. This drives economic growth and social development in the region. However, a few studies have presented conflicting evidence regarding the impact of ports on shaping regional urban dynamics [52]. Meanwhile, Western economies grappled with the unparalleled repercussions of globalization and de-industrialization. Following the financial crisis, unemployment rates remained elevated, economic development was sluggish, and deficits soared to frighteningly high levels. The political atmosphere celebrated the dismantling of barricades and obstacles [53]. Traditional construction methods lead to the development of emerging technologies which in their initial stage (where we currently are) exhibit lower performance due to technical, organizational, and economic challenges, as well as limited integration within an economic environment still primarily influenced by established and traditional technology. Nevertheless, forecasts based on analysis of current patterns indicate that emerging technologies will outperform traditional ones in the long run [54].

3.2 Social Inequality

Professional engineering instruction not only caters to the country's competitive demands, but it also significantly contributes to graduation and valuable societal contributions. Social inequality refers to the systematic disparities in resources, opportunities, and power among different groups in society. It encompasses the process of addressing specific sets of needs and demands put forth by different individuals or groups involved. The concept of professional engineering emerged as a response to the ongoing advancement of intricate systems [55]. Mechatronics showcases diabetes technologies as exemplars of innovation. Insulin pumps are a highly efficient means of administering insulin and have significant advantages in terms of control. The development of an ultra-low-cost insulin pump (ULC) has the potential to significantly increase insulin pump usage. Separating the hardware and computation components results in a device that costs 12–20 times less than currently available traditional-style insulin pumps, while also introducing a new spring-driven design [56]. Figure 2 presents a debate on the social imbalance within the field of mechatronics. Firstly, we need to ensure access to education. Secondly, we should focus on the representation of mechatronics. Next, we should consider career prospects. Lastly, it addresses resource allocation.

Fig. 2 Social inequality of mechatronics

3.2.1 Access to Education

Presently, social, economic, and technological advancements are driving the development of education systems worldwide. Historical evidence demonstrates that the implementation of industrial revolutions led to rapid and significant change. Entrepreneurs transform inventions into marketable innovations, foster the growth of new enterprises, and ultimately respond to consumer demand for novel products and services that enhance their quality of life. Setting in motion the gears of this process leads to a rapid and complete transformation of industry, economics, and society [57]. High-quality Mechatronics education is critical because there is a clear connection between a company's economic strength and the advancement and growth of its young employees. In order to thrive in the global marketplace, a company necessitates highly educated personnel. Engineers are required to collaborate effectively in interdisciplinary teams, often comprising individuals from diverse cultural backgrounds and potentially geographically dispersed [58]. For numerous years, engineering institutions worldwide have been endeavoring to modify the pedagogical approach for educating the upcoming cohort of students. The conventional method of training, which involves a teacher using a blackboard to impart knowledge, has become outdated, has reached its limitations, and has caused some potential customers to feel disconnected. As a result, it is no longer dependable for generating the necessary quantity and caliber of engineers required to push the boundaries of technological advancements [59]. Hence, the impetus behind this endeavor stems from the importance of integrating a comprehensive framework that takes into account emerging demands and combines the advantages of IoT applications with an educational approach to facilitate the swift adaptation of institutions, companies, governments, and society to the rapid proliferation of technology. Thus, this proposal includes the creation and advancement of the subsequent components [60].

3.2.2 Job Opportunity

On a global scale, workers in the manufacturing industry are facing an unprecedented shift in the skills required for their jobs. They must constantly study and improve their abilities in order to stay relevant and maintain their career prospects. The swift incorporation of technology into the workplace, along with other factors like an aging workforce and the shift of job prospects to cheaper labor markets, have adversely affected the job security of contemporary manufacturing workers [61]. The labor market aims to attract individuals who are interested in working in the job opportunities created by corporations and producing institutions (job demand). Therefore, in this market, the employer sector is influenced by supply and demand determinants [62]. Two perspectives approach the implementation of innovation in the mechatronics sector: high-tech business development and the application of mechatronics products in other industries. While there is a substantial body of scientific and statistical literature on the impact of current technology, more scientific studies need to be conducted specifically focused on the use of mechatronics and robotics in business. This makes the topic highly valuable from a scientific standpoint [63]. Automation is an essential component of production systems because it provides a comprehensive view of a fully integrated and engineered system. It symbolizes the interconnection among mechanical, electrical, and manufacturing components. Robots, being integral components of automated systems, serve as a powerful tool to stimulate kids' interest in mathematics, science, and technology. Robotics is a multidisciplinary field that encompasses several fields, such as mechanical and electrical engineering. It involves activities such as design, modeling, sensor integration, signal conditioning, and control systems [64].

3.2.3 Access to Technology

Mechatronics is an interdisciplinary field that integrates electronics, mechanics, and computer science. The digital revolution is fueling the rapid expansion of mechatronics beyond the industrial sector, infiltrating many other industries such as agriculture, healthcare, security, and transportation [65]. Since around 2010, there has been a notable advancement in precision farming technologies. These advancements include the development of affordable and enhanced sensors and servomotors, low-cost microprocessors, cloud-based systems, sophisticated data processing, advanced automation capabilities, and onboard computers. The evolution of Agriculture 4.0 coincided with a comparable advancement in the industrial sector, known as Industry 4.0. The term "Agriculture 4.0" applies this vision to future agricultural output [66].

In recent decades, automation and robots have significantly enhanced the efficiency and cost-effectiveness of industrial production and commodity handling. Comparable transformations have occurred in the agricultural industry in recent years, exemplified by the advent of autonomous tractors and harvesters. Furthermore, GPS and vision-based devices are currently being sold commercially. In recent times, researchers have initiated experiments involving autonomous systems that incorporate field integration [67]. Technological convergence (TC) is a significant aspect of technological development in Industry 4.0. Due to the merger of manufacturing technology and the foundational technologies of Industry 4.0, a new trend of manufacturing innovation is emerging. Manufacturers must adopt emerging technology, such as sophisticated analytics and cyber-physical system-based techniques, to enhance their efficiency and productivity [68]. In order to thrive in a successful career, engineers must continuously acquire new skills and knowledge due to the rapid pace of technological progress. Engineers must possess current knowledge of Industry 4.0 technologies, which include automation and data interchange in manufacturing. These technologies encompass cyber-physical systems, the Internet of Things, cloud technology, and the intelligent factory [68].

3.2.4 Cultural and Behavioral Changes

Agile methodologies primarily originate from the software industry, drawing on its culture, which poses several challenges when applied to the development of mechatronic systems. Nine fundamental concepts form the foundation of ASD (Agile Systems Design), guiding the creation of agile mechatronic systems. By implementing these principles, the development process becomes more agile and tailored to the specific scenario and needs [69]. However, typical Tier 1 suppliers focus on something other than electronic design and software development. In the future, as electric vehicles become more prevalent and have fewer mechanical components, major technology companies will create automotive systems and advanced driver assistance systems (ADAS) that rely on computer vision and artificial intelligence (AI). These systems will aid in the improvement of driving skills and behavior, and the market will offer not only essential products but also high-quality ones with additional features, all at a competitive price [70]. The comprehensive outline approach combines traditional engineering and architectural challenges with emerging technologies like automation, robotics, and information communication technologies (ICT). The analysis begins by examining the impact of emerging technologies on data collection for surveying, inspecting, and monitoring purposes. It takes into account the evidence that the utilization of robotic systems and sensor networks leads to the generation of novel datasets. Digital tools may then process these datasets to construct sophisticated models [71]. The response and management of system designers and educators to these developments and problems will significantly influence the development of both the Internet of Things and mechatronics in the future. This article provides an overview of the connection between the Internet of Things and mechatronics. It examines the difficulties and obstacles that system designers and implementers have when it comes to effectively managing the necessary changes [72].

4 Environmental Impact

The closely related fields of mechanics, electrics and electronics, control engineering, and computer sciences frequently serve as the foundation for mechatronics engineering products. Generally speaking, mechatronic systems fall into two groups. Within the first category, the focus is on shrinking, reducing production costs, and increasing reliability through the spatial integration of mechanics and electronics, with the goal of achieving a high density of mechanical and electronic operations within the available area [73]. For the majority of agricultural tasks, including spraying, managing livestock, harvesting, and post-harvesting, there are both entirely and partially automated systems available. Additionally, current systems use robotics and sensors to implement "precision farming." Modern technology also contributes to workplace health and safety. For example, mechatronic solutions safeguard machinery workers from shearing, tangling, and crushing [74]. Because certain external items have the potential to seriously damage flight vehicle structures, it is necessary to detect any impacts that may occur. For instance, large-scale civil airplane construction frequently employs composite materials. However, internal and hardly noticeable delamination can result from low-velocity impacts like bird attacks, hail, debris, or maintenance instruments [75]. To get these kinds of advantages, product design is essential. Throughout the product's lifecycle, we should assess the product design elements such as material, architecture, shape, and size, as well as functions and modularity, with a focus on its use and eventual disposal. Taking into account the disassembly features is necessary to improve the product's maintainability [76]. Matching hydraulic and electric power, as well as recovering and releasing recovered energy, are critical technologies for electro-hydraulic hybrids, as previously stated. The system is a unique combination of hydraulic and pneumatic energy storage. Researchers discovered that the system's volume and weight had decreased by 92.8% and 83.1%, respectively, making it environmentally friendly [77].

5 Case Studies

Around the world, an increasing number of educational institutions are using technology-based solutions in their operations. Open educational resources are replacing paper books with digital instructional content; laptops, tablets, and mobile

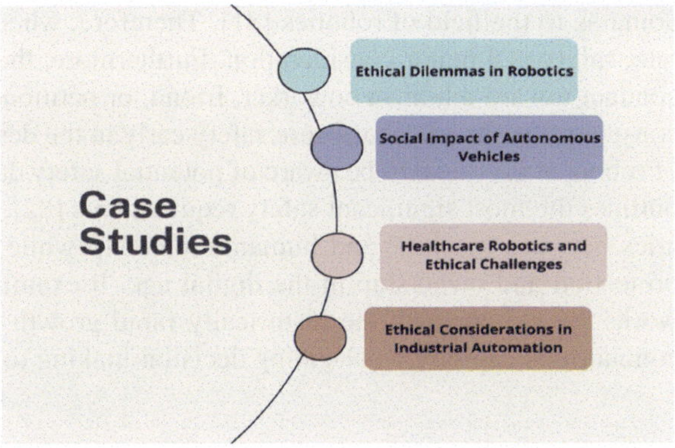

Fig. 3 Case studies of mechatronics

phones are replacing notebooks; interactive whiteboards are replacing analog blackboards; and most importantly, the standardized education process is evolving into more individualized learning opportunities tailored to each student's academic strengths, weaknesses, preferences, and goals [78]. Figure 3 discusses mechatronics case studies with a particular emphasis on ethical issues and ethical concerns in industrial automation, as well as the societal implications of autonomous vehicles, healthcare robotics, and ethical dilemmas in robotics.

5.1 Ethical Dilemmas in Robotics

The potential effects of upcoming robots and artificial intelligence systems have drawn more attention in recent years. Well-known intellectuals have expressed concern in public about the possibility of a dystopian future if these systems continue to grow more complicated. These cautions contrast with the most advanced robotics and artificial intelligence technologies available today. This article scrutinizes research that explores the potential of robots and artificial intelligence systems, along with the ethical measures required to avert a gloomy future [79]. Rescue robot operators can reach challenging locations that are either unreachable, too dangerous, or too slow for people to enter. They can also act as platforms for remote sensing, enabling communication between people and the devastated environment. For instance, a rescue robot can aid in the visual inspection and mapping of a collapsed building's interior, evaluate the extent of damage, and install seismic, thermal, or acoustic sensors to monitor the situation [80]. Robotics, following information and communication technology (ICT), biotechnology, nanotechnology, and neuroscience-related technologies, is gaining regulatory attention as the next significant area of technological advancement. Remark 1 The terms "soft bots," "robotics," "nanobots," and "neurorobotics," used to characterize robots or specific aspects of their design, illustrate the diverse contributions of these earlier, broad technological domains to the field of robotics [81]. Therefore, when humans and robots collaborate, safety is a major consideration. Furthermore, the ethical component and their conduct toward a human coworker, friend, or petitioner are increasingly taken into consideration. In order to ensure safety early in the design and development process of robots, it is crucial to be aware of potential safety dangers. Collaborative robotics outlines the most significant safety requirements [82]. It negotiates the hazy boundaries between machine and human authorship while posing important queries on protection and ownership in the digital age. It examines how current legal frameworks are coping with the historically rapid growth of technology and the moral conundrums caused by delegating decision-making to intelligent algorithms [83].

5.2 Social Impact of Autonomous Vehicles

One of the most eagerly awaited technological advancements of our day is autonomous driving, which could have profound social ramifications. When social scientists talk about self-driving cars, they usually focus on the most basic utilitarian issues like efficiency, safety, and regulation. On the other hand, when people talk about self-driving cars in public, they usually support a basic form of technical determinism [84]. Road-based transportation is considered here, which includes the integration of self-driving technology into both private vehicles and public transportation options, despite the technology's adaptability to a wide range of vehicle types. The literature now in publication frequently uses the terms "autonomous," "self-driving," and "driverless" synonymously in this context to differentiate them from CVs (conventional vehicles) [85].

Autonomous vehicles (AVs) must be able to handle complex observations and interactions in real-world scenarios, like merging onto highways and making unprotected left-hand turns, which are difficult even for human drivers. Since AVs do not yet comprehend human behavior, cautious behavior is necessary for safe functioning. Conservative driving, particularly near intersections, causes traffic bottlenecks [86]. The use of autonomous vehicles (AVs) brings up significant ethical and social issues, particularly those pertaining to justice, accountability, and dignity. The focus has been on the particular issues raised by the way AV technology will impact the lives and careers of semi-professional and professional drivers. We aim to comprehend these ethical and societal difficulties from the views of the drivers themselves, in contrast to prior studies of similar concerns that have concentrated on the opinions of experts [87]. Figure 4 depicts the temporal distribution of vehicles that remain on the route. It has observed the highest traffic at about 6 p.m. and the lowest traffic at midnight, as depicted in the figure. The peak traffic hours often fall between 8 a.m. and 2 p.m.

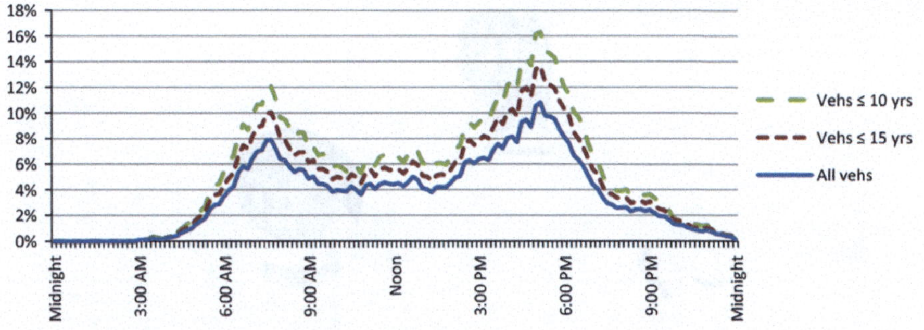

Fig. 4 Analysis of car usage based on time of day and vehicle age (Puget Sound Regional Council, 2006) [88]

5.3 Healthcare Robotics and Ethical Challenges

The utilization of robots in the healthcare sector provides promising prospects for assisting a substantial population. Robots can assist individuals with cognitive, sensory, and motor disabilities, provide aid to those who are sick or injured, offer support to caregivers, and assist clinical personnel. Additionally, it delineates many significant technological, logistical, and design obstacles encountered [89]. These reflections on the future of healthcare involving machines provide valuable insights into ethical considerations, allowing us to examine our concerns, formulate more sophisticated arguments regarding potential issues, and gain a deeper understanding of the current advancements in healthcare, as well as the broader progress in robotics and our technological society [90].

Figure 5 shows that Care-O-Bot 4, developed by Fraunhofer, is a versatile service robot that can perform many tasks and fulfill different roles. It may function as a mobile information center, collect and transport items, and serve as a security or surveillance tool. Additionally, its open-source software interfaces make it suitable for use as a research platform. The system comprises an omnidirectional platform, a torso with a sensor ring positioned above it, and a head that has a touch screen, as depicted in Fig. 5 [92]. The system explores the potential applications and benefits of artificial intelligence (AI), while also addressing the specific ethical challenges that arise in some cutting-edge industries. It also explores possible ethical governance structures that can address these concerns. Frontier AI research and development in medical care have significant potential, but the ethical problems arising from their uses have necessitated additional governance standards. In order to ensure the reliability of AI applications in healthcare and medicine, there have been suggestions to establish an ethical global governance structure and system, along with specific standards [93]. This study scrutinizes the practical and ethical challenges associated with developing and evaluating social robots that aim to establish security and earn the trust of their human users. Trust is of utmost importance in collaborative scenarios where humans must accept information from the robot and comply with

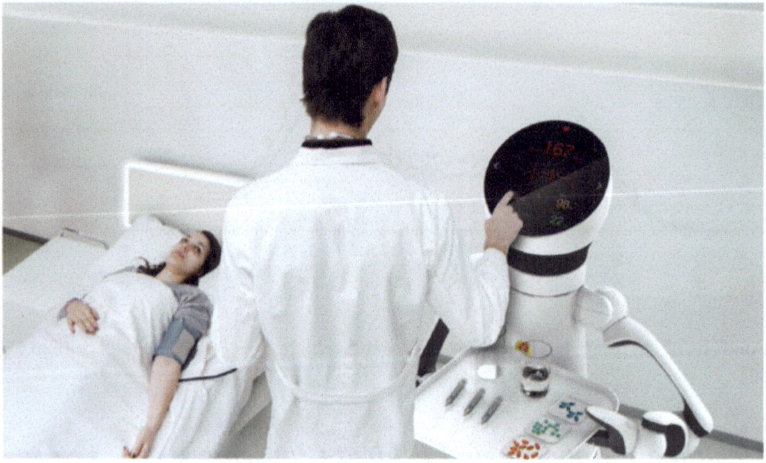

Fig. 5 Care-O-bot 4 [91]

its ideas. Persuasiveness in these situations closely correlates with trust [94]. Therefore, it is necessary to apply the ethical concepts of nonmaleficence, beneficence, autonomy, and fairness to robotics. Improved comprehensive answers and solutions to all ethical challenges in robotics remain a priority. Establishing ethical guidelines for robot deployment and conducting research on social service robots (SSR) are crucial to integrate ethics into interactions with socially assistive robots [95]. However, the shifting population dynamics in affluent nations present growing difficulties in providing care and assistance to older people. As life expectancy increases and birth rates decline, the proportion of elderly individuals relative to younger ones grows, resulting in a greater number of people requiring healthcare [96]. The lack of literature on the ethical aspects of artificial intelligence (AI) in low- and middle-income countries (LMICs), as well as in the field of public health, highlights the urgent requirement for additional study on the ethical implications of AI in both global and public health. This research is necessary to guarantee that the development and use of AI are ethical and inclusive for all individuals, regardless of their location [97].

5.4 Ethical Considerations in Industrial Automation

As a result of significant advances in information and communications technology, the manufacturing industry is facing revolutionary changes whereby production

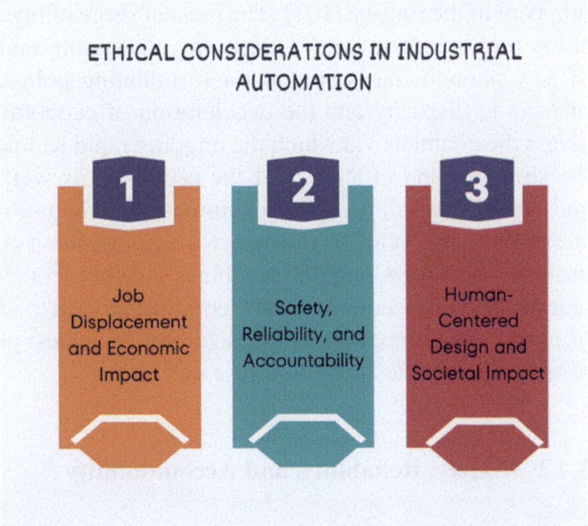

Fig. 6 The types of ethical considerations in industrial automation

processes will become increasingly digitized and interconnected through cyber-physical systems. A vital component of these new complex systems will be intelligent automation and human–robot collaboration. Manufacturing systems have traditionally segregated industrial robots from people due to the dangers posed by their operational speeds and heavy payloads [98].

Figure 6 explores the ethical considerations in industrial automation, encompassing job displacement and economic impact, safety, reliability, and accountability, as well as human-centered design and societal impact. Industrial automation is necessary to ensure the stability of work ethics in any industry.

5.4.1 Job Displacement and Economic Impact

Companies are quickly using new technology to automate their workplaces, such as driverless cargo trucks, artificially intelligent mortgage approvals, machine-learning-based paralegals, and algorithmic supervisors. Technological progress raises a number of questions for both the commercial world and society. More advanced engineering cannot solve this issue, as Thomas Donaldson recently stated [99]. Theoretically, robots have the potential to decrease employment and salaries. By analyzing the variation in exposure to robots, it is possible to determine the local effects of robots on industry-level robotics advancements and employment levels in local industries. Our analysis indicates significant and adverse impacts of robots on both employment levels and earnings within commuting zones [100]. Industries use different technological feasibility measures, referring to the proportion of work that robots can replace, to determine the effect of robot utilization on employment. The statistics enable us to distinguish between effects based on the level of employment intensity in the routine [101]. The present surge of invention carries implications that elude traditional economic reasoning. Assessing and forecasting the implications of new phenomena is crucial for formulating policies that mitigate the widening of income disparity and the deceleration of economic growth. The objective is to assess the channels via which the ongoing rapid technological advancements impact the structure and efficiency of the economy, as well as the actions of individuals and organizations [102]. Firm automation is the use of technology and software to streamline and optimize various processes within a company. In this context, I will provide a new paradigm that combines stakeholder theory and social contract theory. Our methodology utilizes these theoretical models to identify the ethical implications of corporate automation, emphasize best practices, provide recommendations, and reveal opportunities for future research [103].

5.4.2 Safety, Reliability, and Accountability

For instance, the use of heavy machinery and the handling of dangerous substances classify New York factories as safety–critical. Therefore, certain assurances offered by the "things," such as those pertaining to performance and availability, are considered essential to safeguarding the production processes and the resultant products [104]. The Human-Centered Artificial Intelligence (HCAI) framework guides designing systems that balance human control and computer automation to enhance human performance. It helps identify scenarios where either complete human control or complete computer control is required while also preventing the risks associated with excessive control from either humans or computers. The techniques employed in HCAI are highly inclined to generate designs that exhibit qualities of reliability, safety, and trustworthiness (RST) [105]. The distribution system operators need help with centralized control and integrated functioning, particularly in ensuring safety and reliability. In order to address these challenges, the concept of "distribution system automation" emerged, utilizing information and communications technology to seamlessly integrate all the essential components of a distribution system [106]. Many existing works explore the technology, uses, and difficulties of the prior industrial standard known as Industry 4.0, as well as the accompanying industrial technologies. For instance, the research examines the enabling technologies used in the industry standard and the rationale behind their integration into the standard [107]. Management strategies that include leadership dedication to safety, safety-focused recruitment and training, comprehensive reporting of failures and near misses, internal review boards for addressing issues and formulating plans, and adherence to industry standard practices all contribute to the establishment of a safety culture within organizations. Government interventions and regulations, external audits by accounting firms, insurance companies offering compensation for failures, non-governmental and civil society organizations promoting design principles, and professional organizations and research institutes establishing standards, policies, and innovative ideas all contribute to the trustworthiness certification [108].

5.4.3 Human-Centered Design and Societal Impact

The term "human-centered" is frequently employed in a broad sense without any specific dedication to a philosophy or comprehensive methodology. This text presents several philosophical perspectives and multiple human-centered design techniques, examining their impact on the advancement of artificial intelligence. This text asserts that humanistic design research should have a crucial role in collaborating with technologists and politicians from other disciplines to reduce the adverse effects of AI [109]. The framework consists of three fundamental tiers. At its core, the contact between the user and the product elicits three distinct types of direct product experience: the aesthetic experience, the experience of meaning, and the emotional experience. The second level encompasses more indirect and long-term forms of influence, specifically on behaviors, attitudes, overall experiences, and the knowledge of users and stakeholders. The third and ultimate level symbolizes the overall standard of living and societal conditions [110]. Human-centered design (HCD) can be a crucial tool in addressing today's intricate care concerns. HCD's systemic and compassionate approach, combined with its emphasis on creativity, allows for effective change. HCD, or human-centered design, focuses on identifying human needs in order to create goods or services that fulfill those requirements. The finished design is comprehensible and functional, thereby achieving the intended objectives and providing a meaningful and enjoyable user experience [111]. Many projects fail because they make basic assumptions about what end users desire and do, or because implementing them on a broad scale is much more complicated than conducting small trials. As digital technologies apply, the outcomes continue to evolve, as does the delivery method. The rationale for frequent changes in health programs is not only due to quick technological advancements but also because stakeholders often reasonably request modifications to incorporate numerous health programs or adapt to local infrastructure and health professional routines [112]. The most important parts of this approach are (a) figuring out what factors affect implementation at different levels by understanding the users' needs and situations; (b) adapting implementation strategies to each situation by redesigning interventions as needed; and (c) evaluating the processes and results of implementation, such as how the main focus on human-centered design (HCD) usability is connected to other important factors like acceptability, feasibility, and appropriateness [113].

6 Ethical Design and Responsible Innovation

In creating mechatronic systems, ethical design entails considering society's impact, reducing harm, and fostering justice. At every stage of the technology lifecycle, responsible innovation aims to minimize risks and disparities while promoting sustainability, responsibility, and transparency for the benefit of society [114]. All parts of Fig. 7 work together to make a single picture of responsible innovation and ethical design in mechatronics. This shows how ethical principles and innovation

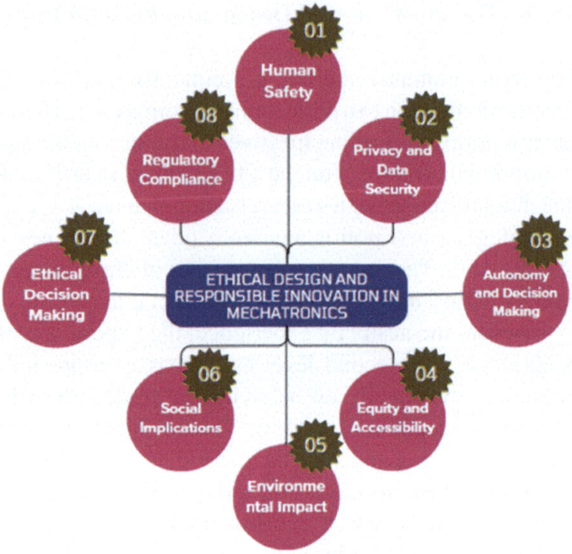

Fig. 7 Ethical design and responsible innovation in mechatronics

techniques are connected when making mechatronic systems that do the most good and the least harm.

6.1 Designing Ethical Mechatronic Systems

The development of ethical mechatronic systems must place a high priority on inclusion, privacy, and human safety. To preserve moral principles and encourage responsible use, it is necessary to consider the broader societal ramifications and ensure that decision-making procedures are open and transparent [115]. These days, creating novel and interdisciplinary goods such as mechatronic systems calls for an integrated strategy incorporating various viewpoints and engineering, design, and ethics expertise. It is important to redesign or modify the method of developing new products to foster teamwork, flexibility, and continuous education. This will allow teams to effectively handle ethical and societal issues while navigating intricate technological problems. This means embracing interdisciplinary cooperation, iterative design techniques, and stakeholder participation throughout the development lifecycle [51].

6.2 Ethics in CPHS

Deployment is still a way off. Scholars such as McBride (2016), Goodall (2016), and Luetge (2017) have already written numerous articles addressing a variety of ethical concerns in this field. As self-driving cars become more widely used and incorporated into society, ethical concerns ranging from safety and accountability to replacing human drivers will become more significant and prominent. Numerous questions arise as we look to the future and as this technology reaches the power stage of application. The following queries from the Ethics Commission of the Federal Ministry of Transport and Digital Infrastructure in Germany's report on the ethics of driverless and connected cars provide a peek into this future [5]. Summary: Fig. 8 shows Cyber-Physical-Human Systems (CPHS), a set of moral guidelines for dealing with the problems that come up when digital, analog, and human parts are combined. The above system must be humane, adhering to ethical norms and human rights while also considering privacy, justice, safety, and transparency. These elements work together to create systems that serve society as a whole. On the other hand, in the realm of mechatronic systems, human–robot interaction (HRI) has become a vital subject of study with great potential for improving many aspects of human life and pushing technological capabilities. It starts by summarizing the significance of HRI and how it affects the advancement of mechatronic systems [116]. The cyber-physical-human systems (CPHS) domains are looked at and put into groups in Fig. 9. The focus is on how they can be used in different fields, like aerospace, robotics, healthcare, automotive, power systems, and the process industries. Four categories—human-in-the-plant, human-in-the-controller, human–machine control symbiosis, and humans-in-multiagent loops—are emphasized for human-in-the-loop control using architectural templates. Each case's technological evolution and a timeline of advancements are described [117–120].

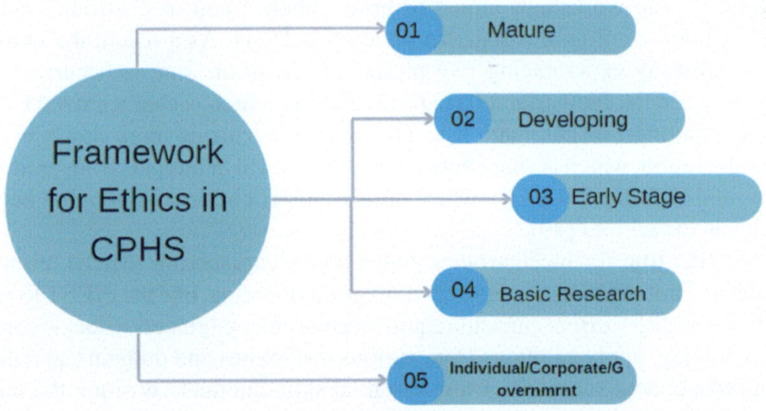

Fig. 8 Summary of framework for ethics in CPHS

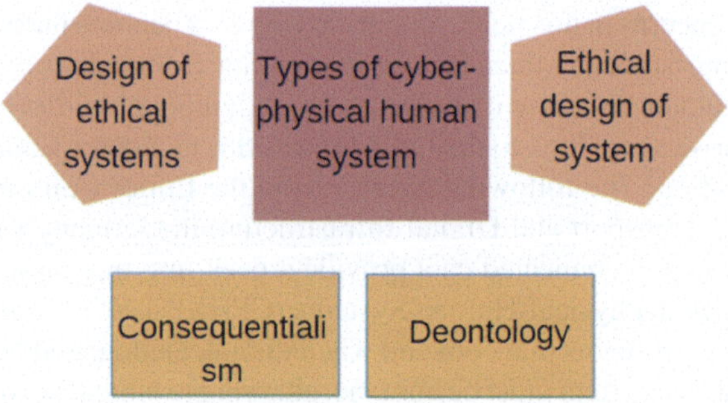

Fig. 9 Types of CPHS (physical human system)

6.3 *Incorporating Ethical Considerations in Engineering Education*

Including ethical theory and case studies in the curriculum, encouraging students to apply ethical principles in their design and decision-making processes, and promoting critical thinking about the impact of technology on society are all important ways to incorporate ethical considerations into the teaching of mechatronic engineering. This will prepare upcoming engineers to handle difficult moral decisions and prioritize human values. In September 2019, the fourth and final workshop on the Future of Mechatronics and Robotics Education (FoMRE) took place at Lawrence Technological University in Southfield, Michigan. Faculty from multiple universities coordinated this workshop with funding from the National Science Foundation and industrial partners. The workshops aimed to bring robotics and mechatronics courses, minors, and degree programs together cohesively [121]. As a result, the industrial world is currently experiencing two phases of transition: one is occurring more quickly between the Industrial 3.0 (I3.0) revolution, which is characterized by electronics, computing, and automation, and the other is occurring more slowly between the I4.0 revolution, which is characterized by the removal of humans from production systems, and Industry 5.0 (I5.0), which aims to foster closer collaboration between operators and machines [49].

This is also true for mechatronics and robotics engineering (MRE), even with focused K–12 initiatives to promote diversity in engineering, like the FIRST Robotics Competition [122]. Current curricula equip mechatronics engineers to tackle complex problem-solving, apply solutions from various disciplines and domains, and deliver domain-independent solutions. These essential skills uniquely position the mechatronics engineer to embrace sustainable development [123]. A culture that prioritizes profit maximization has completely globalized and populated the real world [124]. Engineering developments in computer technology and embedded system design, engineering design has seen a transformation in recent years. It is uncommon for engineers to create mechanical designs without also including electrical and electronic components. Mechanical engineers and mechanical engineering technologists must possess a multidisciplinary understanding of both mechanical and electrical systems.

Undergraduate engineering technology programs specifically include mechatronics courses in their curriculum. The design process known as mechatronics is multidisciplinary in nature, combining concepts from several technical specialties, such as controls, electrical, and mechanical engineering [125]. The real world develops a variety of hands-on laboratories that cover numerous topics related to mechatronics, microcontrollers, communications, and control. The organization of the practicals enables students to tackle real-world situations in a systematic manner. The program's most significant aspect is its incorporation of multidisciplinary knowledge together with the professional and technical abilities necessary to pursue a successful career [126]. However, the primary reason for this difficulty is the need for integration, teamwork, and holistic approaches during the design process [127]. Mainly, performance evaluation and quality standards are quickly becoming indispensable instruments for designing and evaluating educational initiatives [9, 128].

6.4 Responsible Research and Innovation Practices

In mechatronics, responsible research and innovation (RRI) techniques mean including social, ethical, and environmental factors at every stage of the study process. This includes involving stakeholders, foreseeing possible effects, and proactively resolving moral conundrums in order to guarantee that technical innovations serve society while reducing harm and fostering sustainability [29, 129]. This research objective also encompasses the following figures: (i) The search for "mechatronic" in the traditional indexing fields of title, keywords, and summary yields a range of publications in the central databases Web of Science and Scopus, totaling 33 thousand documents. (ii) Since we think that innovation is both a quality and an asset in mechatronics, when we search for "innovation*," we get 854 publications, which means that the sample proposal can be used for the bibliometric, systematic review, and qualitative methods that were planned [51, 127]. Mechatronics engineering is a demanding field that requires an approach and methodology distinct from traditional engineering [63]. Businesses use a range of business process optimization technologies to help with planning, development, delivery, hiring new staff, and creative work. To meet the strategic goals of a business, business process optimization must produce a number of outputs, including waste avoidance, error correction, cost reduction, and improved departmental performance. Business process management (BPM), business process automation, and robotic process automation (RPA) are the three methods of business process optimization, according to Haley Burton (2019). Every technique describes how it improves an organization's competitive edge by improving multiple processes and reducing repetitive or inefficient work [130, 131]. In the given context, object control employs a model as an abstract depiction of a system. These models occasionally include the environment or the area surrounding the system under study. The idea of a model is closely associated with the so-called metamodel, or model of a model. It functions as a limited semantic notation or language that describes the model. Metamodels are a long-standing tool in the engineering community and can be textual or graphical. However, until recently, their only function was to describe and comprehend a variety of environmental events. Their development is now a critical stage in control system design [132, 133].

7 Regulatory and Policy Perspectives

The creation of frameworks that control the development, application, and use of mechatronic systems while guaranteeing the preservation of ethical, private, and safety concerns is the focus of regulatory and policy views in mechatronics. To foster innovation and safeguard societal interests and values, we need to establish rules, norms, and oversight mechanisms to address emerging issues such as cybersecurity, data protection, and the ethical implications of autonomous systems [134]. Figure 10 summarizes regulation and policy, offering an analysis of viewpoints on mechatronics-related rules and policies. Safety requirements, regulatory compliance, data privacy security, and so forth are included.

7.1 International Standards and Guidelines

The Association for Molecular Pathology convened a multidisciplinary working group with liaison representation to evaluate the current state of next-generation sequencing-based cancer testing. The group's objective was to establish standardized consensus classification, annotation, interpretation, and reporting conventions for somatic sequence variants [135]. The meeting, held until April 2020, extensively addressed several new topics. New stimulation devices and pulse configurations, the

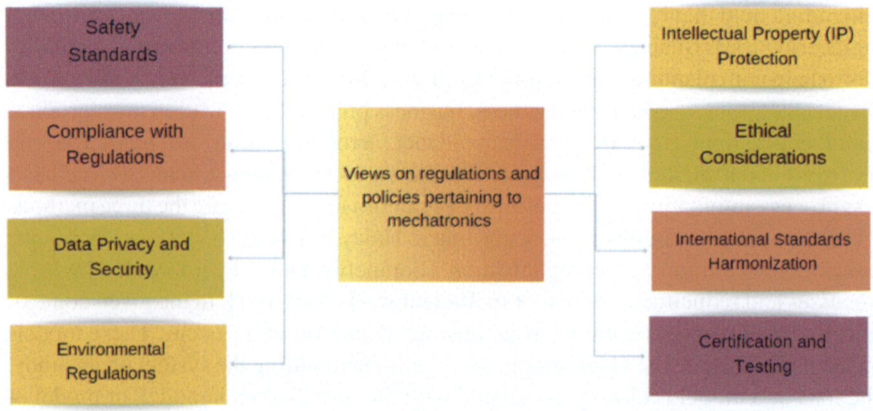

Fig. 10 Views on regulations and policies pertaining to mechatronics

duties and responsibilities of device makers, new uses of transcranial magnetic stimulation (TMS) in neuroimaging and TMS guided by imaging or robots, the combination of TMS with transcranial electrical stimulation, making sure safety during paired associative stimulation interventions, and the risks of using TMS to cause therapeutic seizures (magnetic seizure therapy) are some of the things that worry people about safety [136]. In order to be efficient and valid regulators, RSS systems necessitate specific skills such as knowledge, operational capability, autonomy, and representativeness. Collaboration among business, civic society, and state actors is crucial in order to bring together a whole range of skills and abilities. However, the majority of RSS systems typically only include one sort of actor, which means they need to possess crucial abilities. It investigated the trend from a distributive bargaining perspective [137]. These standards and regulations categorize the test procedures based on the specific abuse conditions that are relevant. Examine the test parameters and conditions used in the test procedures, with a focus on identifying similarities and contrasts. The list includes pertinent upcoming standards and regulations [138]. Communities, scientists, policymakers, and land managers participate in this solution-focused strategy. Its goal is to restore ecological damage and establish a more beneficial connection between humans and the natural world. Combining ecological restoration with conservation and sustainable use is the key to transform the constantly deteriorating local, regional, and global environment into one that is generally improving [139]. The panel determined that a narrow definition of synbiotics as a mere combination of probiotics and prebiotics could hinder the development of synbiotics specifically tailored to work collaboratively. Imposing the condition that each component must independently satisfy the proof and dose criteria for probiotics and prebiotics could also be a challenge [140]. Currently, we lack the necessary technology and infrastructure to create large-scale immersive virtual worlds accessible across multiple platforms. However, researchers are increasingly studying the profound effects of the metaverse. The sectors affected by this include marketing, education, healthcare, and the societal effects related to factors of social interaction resulting from widespread adoption. Additionally, there are concerns regarding trust, privacy, bias, disinformation, and the application of law, as well as psychological aspects associated with addiction and its impact on vulnerable individuals [141].

7.2 *Government Regulations and Oversight*

With the increasing influence of algorithms in government decision-making globally, officials have been discussing strategies to maximize the advantages of algorithms while minimizing their negative impacts. Requiring human review of algorithmic decisions has emerged as a central element in global efforts to regulate government algorithms. Although there is a widespread shift toward human monitoring, these regulations are based on an unquestioned assumption [142]. The two main objectives of government regulations and oversight are the protection of the public interest

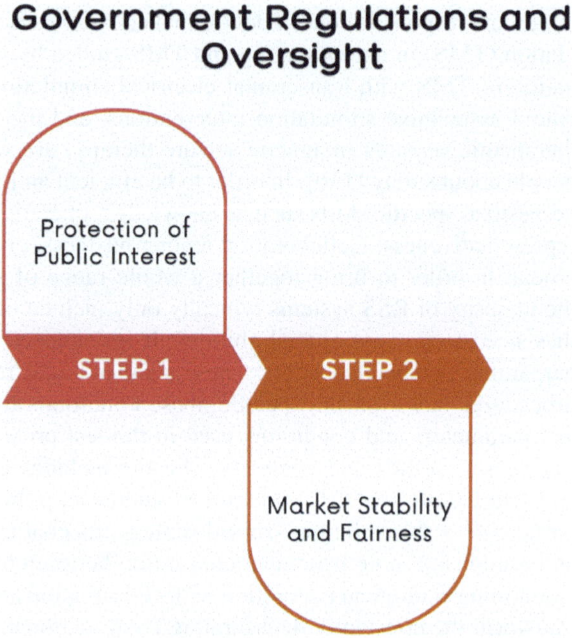

Fig. 11 Government regulations and oversight

and market stability and fairness. Figure 11 presents government regulations and oversight, illustrating the role of government laws and oversight in safeguarding the public interest and ensuring market stability and fairness.

7.3 Protection of Public Interest

Our daily lives now incorporate AI and machine learning devices. What was once considered science fiction, such as self-driving cars and medical robots in Star Wars, has now become a reality. However, it is important to note that movies often fail to mention that these gadgets require extensive programming and training using large quantities of personal data. The primary concern is the inability to consistently obtain this data directly from individuals with explicit agreement. As a result, the development and study of these gadgets primarily focus on the use of personal data for secondary purposes [143]. The regulation restricts personal data processing in certain industries while allowing it in others. The regulation asserts that data processing should be purposefully meant to benefit humanity. The Regulation, while not explicitly defining the specific meaning, suggests the adoption of a balanced approach to safeguard personal data [144]. Concurrently, digital technologies with weaker data privacy safeguards allow for greater data collection. In summary, this evaluation indicates that a significant number of apps adhere to data privacy rules. However, there is still room for optimizing their value in terms of public interest [145]. The

Universal Declaration of Human Rights and Protection of Public Interest defines this freedom not only as the right to express one's opinion, but also as the right to seek, receive, and share knowledge. Whistleblowing refers to when individuals who are or were part of an organization reveal illegal, immoral, or illegitimate activities that their employers are carrying out to others or organizations that have the power to take action [146].

7.4 Market Stability and Fairness

The enduring presence of racial disparities in the labor market, despite the existence of statutory equal opportunities, is a concerning issue that has important implications for the development of hiring practices. In order to tackle these inequalities, we propose implementing a two-tier labor market system. The first tier is a temporary labor market, where firms are required to hire workers in a way that ensures equal representation of different groups. The second tier is a permanent labor market, where firms have the freedom to hire the most qualified candidates as they see fit [147]. Fair value measurement pertains to forecasts that are more widely spread. Implementing the measurement-basis information mandated by SFAS 157 leads to more precise predictions. Still, it has no significant impact on banks that have a substantial share of their assets valued at fair market value. Furthermore, measurement is associated with improved forecast accuracy, whereas measurement is associated with higher forecast dispersion [148]. Regression models excel in handling numerical and continuous values. Ensemble machine learning approaches, particularly when combined with weak learners such as decision trees, are effective in dealing with larger datasets [149]. A sequence of eight controlled experiments conducted in a laboratory setting resolved this matter. The findings indicate that fairness can influence market pricing, although the effects may diminish over time, which has significant implications for market stability and fairness [150].

8 Ethical Codes of Conduct for Mechatronics Professionals

The curriculum's primary goal is to enable students to acquire fundamental knowledge of embedded systems and robotics and effectively implement them. In order to fulfill this objective, a range of hands-on laboratory exercises will be used to instruct on various aspects of communications, control systems, mechatronics, and microcontrollers. The structure of the practicals enables students to methodically address and resolve real-life challenges [126]. Currently, the advancement of science involves a dual process: firstly, there is the derivation of sciences, which is linked to their growing specialization; secondly, new studies emerge as a result of the mutual penetration or symbiosis of diverse domains of knowledge. The fields of medical and technical sciences provide an exemplary illustration. Indeed, an examination of the most recent advancements in this sector demonstrates the infeasibility of independent progress in these domains [151]. During the model establishment, professionals score the index system using the analytic Ethical Codes of Conduct approach to determine the weight of each index. Finally, the evaluation result is calculated. Subsequently, we utilize the findings above to calculate a score, denoted as P, which assesses the effectiveness and rationality of the organizational human resource professional norms system. This evaluation aims to offer recommendations and serve as a reference for the development of an exceptional professional norm system [152]. The Ethical Codes of Conduct for Mechatronics Professionals encompass several fundamental principles, namely safety, responsibility, integrity, sustainability, and continuous learning. Figure 12 illustrates the five significant issues that have a substantial impact on the Ethical Codes of Conduct for Mechatronics Professionals. The safety aspect plays a significant role in the implementation of mechatronics. Following that, responsibility becomes an essential element in mechatronics professional practice. Next, integrity plays a crucial role in adhering to ethical codes. Subsequently, sustainability is emphasized. Lastly, continuous learning is essential for mechatronics professionals to uphold ethical codes of conduct.

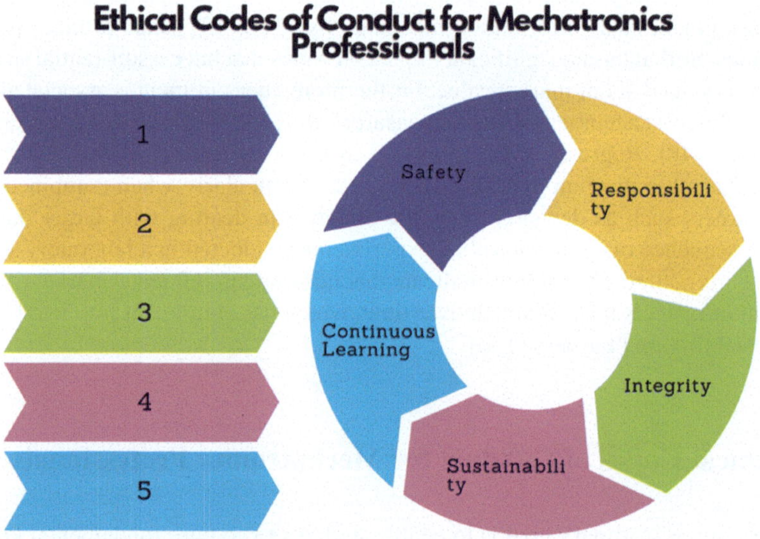

Fig. 12 Key points of ethical codes of conduct for mechatronics professionals

9 Conclusion

The ethical and social aspects of mechatronics are of utmost importance in ensuring that technical breakthroughs have a positive impact on society. Using moral theories like utilitarianism, deontology, virtue ethics, and rights-based ethics together provides a strong foundation for guiding the development and use of mechatronic technology. These frameworks ensure that innovations not only enhance technical skills but also comply with ideals of safety, privacy, sustainability, and human dignity. By giving priority to these ethical standards, engineers and developers can cultivate trust and faith in mechatronic systems, thereby enabling their advantageous incorporation into everyday life. The chapter emphasizes the importance of ongoing assessment and open decision-making procedures in order to synchronize technological advancements with ethical obligations, with the ultimate goal of improving the overall standard of living and fairly addressing social needs.

Moreover, the chapter emphasizes the significant societal consequences of mechatronics, recognizing both the possible disruptions and advantages it offers. The economic consequences, such as the displacement of jobs and the requirement for acquiring new skills, highlight the importance of taking proactive steps to guarantee fair access to education and retraining. On the other hand, mechatronics provides substantial advantages, such as the generation of fresh job prospects and the advancement of technical availability. Case studies exemplify the intricate ethical dilemmas presented by autonomous vehicles and healthcare robotics. These case studies emphasize the significance of human-centered design in effectively tackling these obstacles. The chapter also highlights the crucial importance of regulatory frameworks, international standards, and academic institutions in promoting a culture of responsible innovation. By following these standards and fostering ethical consciousness, the mechatronics industry can attain a harmonious equilibrium between technological progress and societal welfare, guaranteeing that its contributions are fair and enduring.

References

1. Adegbite AO, Nwasike CN, Nwaobia NK, Gidiagba JO, Enabor OT, Dawodu SO, Ezeigweneme CA (2023) Mechatronics in modern industrial applications: delving into the integration of electronics, mechanics, and informatics. World J Adv Res Rev 20(3)
2. Khan MR, Antong H, Wahid AN, Nor KAM, Hamid SBA, Sophian A (2021) Mechatronics engineering curriculum in the new perspective. Engineering Professional Ethics and Education 2021 (ICEPEE'21), p 80
3. Westerlund M (2020) An ethical framework for smart robots. Technol Innov Manage Rev 10(1)
4. Wang H, Khajepour A, Cao D, Liu T (2020) Ethical decision making in autonomous vehicles: challenges and research progress. IEEE Intell Transp Syst Mag 14(1):6–17
5. Khargonekar PP, Sampath M (2020) A framework for ethics in cyber-physical-human systems. IFAC-PapersOnLine 53(2):17008–17015

6. Braun BF, Cordts M, Siebeneich D (2022) Fairness, accountability, transparency, and explainability in intelligent automation. cii Student Papers 2022, 1
7. White R, Bradley D (2022) Artificial intelligence, ethics and privacy. EcoMechatronics: challenges for evolution, development and sustainability. Springer International Publishing, Cham, pp 129–152
8. Maksuti R, Osmani H, Ismaili N, Sulejmani K, Mehmedi I (2022) Development of mechatronic technology through a transdisciplinary knowledge approach. J Appl Sci-SUT (JAS-SUT) 8
9. McFall KS, Huang K, Gilbert HB, Jouaneh MK, Bai H, Auslander DM (2020) Mechatronics and robotics education: standardizing foundational key concepts
10. Venugopal P, Priya SA, Manupati VK, Varela MLR, Machado J, Putnik GD (2018) An analysis of the impact of UTAUT predictors on the intention and usage of electronic health records and telemedicine from the perspective of clinical staffs. Int J Mechatronics Appl Mech 4:263–269
11. Vicković B, Popić S, Jovanović MD, Rodić A (2017) Artistic design of the customized robot with environmental, social and cultural impacts to human society. In: Advances in robot design and intelligent control: proceedings of the 25th conference on robotics in Alpe-Adria-Danube region (RAAD16). Springer International Publishing, pp 593–602
12. Iizuka ES, de Moraes GHSM, de Souza MG (2024) College environment and entrepreneurial intention in high school. Revista de Gestão 31(1):101–114
13. Bibri SE (2022) The social shaping of the metaverse as an alternative to the imaginaries of data-driven smart cities: a study in science, technology, and society. Smart Cities 5(3):832–874
14. Kemper S (2021) Locating creativity in differing approaches to musical robotics. Front Robot AI 8:647028
15. Samak C, Samak T, Krovi V (2023) Towards mechatronics approach of system design, verification and validation for autonomous vehicles. In: 2023 IEEE/ASME international conference on advanced intelligent mechatronics (AIM). IEEE, pp 1208–1213
16. Janschek K, Morozov A (2015) Dependability aspects of model-based systems design for mechatronic systems. In: 2015 IEEE international conference on mechatronics (ICM). IEEE, pp 15–22
17. Portillo-Velez RDJ, Vázquez-Santacruz E, Morales-Cruz C, Gamboa-Zuniga M (2016) Mechatronic design and manufacturing of an affordable healthcare robotic bed. J Rehabil Assistive Technol Eng 3:2055668316668792
18. Damelang A, Haupt A, Abraham M (2018) Economic consequences of occupational deregulation: natural experiment in the German crafts. Acta Sociologica 61(1):34–49
19. Prathibha S, Palanikumar K, Ponshanmugakumar A, Kumar MR (2024) Application of augmented reality and virtual reality technologies for maintenance and repair of automobile and mechanical equipment. In: Machine intelligence in mechanical engineering. Academic Press, pp 63–89
20. Gu S, Zhang L, Guo S, Zheng L, An R, Jiang T, Xiong A (2022) Communication and cooperation for spherical underwater robots by using acoustic transmission. IEEE/ASME Trans Mechatron 28(1):292–301
21. Trentesaux D, Karnouskos S (2022) Engineering ethical behaviors in autonomous industrial cyber-physical human systems. Cogn Technol Work 24(1):113–126
22. Van de Poel I, Sand M (2021) Varieties of responsibility: two problems of responsible innovation. Synthese 198(Suppl 19):4769–4787
23. van Grunsven J, Stone T, Marin L (2024) Fostering responsible anticipation in engineering ethics education: how a multi-disciplinary enrichment of the responsible innovation framework can help. Eur J Eng Educ 49(2):283–298
24. Voegtlin C, Scherer AG (2017) Responsible innovation and the innovation of responsibility: governing sustainable development in a globalized world. J Bus Ethics 143:227–243
25. Sharma M, Gupta S, Mahesh T (2024) Redefining financial frontiers: ethical imperatives and regulatory dynamics in robotic investment platforms. In: Proceedings of the 1st international conference on artificial intelligence, communication, IoT, data engineering and security, IACIDS 2023, 23–25 November 2023, Lavasa, Pune, India

26. Tokarev A, Berestov A, Klimova S, Tolstov M, Pankratova A (2021) Application of VR instruments in preprofessional education in the area of mechatronics and robotics in a nuclear research university. Procedia Comput Sci 190:745–749
27. Sandhu H, Gangacharyulu D, Agrawal VP (2018) Coding, evaluation, comparison, ranking and optimal selection of nanoparticles with heat transfer fluids for thermal systems. Part Sci Technol 36(1):50–60
28. Adeleke AK, Olu-lawal KA, Montero DJP, Olajiga OK, Ani EC (2024) The intersection of mechatronics and precision engineering: synergies and future directions. Int J Sci Res Arch 11(1):2356–2364
29. Cintra Faria AC, Barbalho SCM (2023) Mechatronics: a study on its scientific constitution and association with innovative products. Appl Syst Innov 6(4):72
30. Rassõlkin A, Vaimann T, Kudelina K, Demidova GL, Lukichev DV, Perepelkina SY (2020) Modem mechatronics and robotics education program: border cooperation between Estonia and Russia. In: 2020 IEEE global engineering education conference (EDUCON). IEEE, pp 1272–1277
31. Rojas J, Carranza K (2024) Check for updates the medical device industry in a Kaizen Environment in the year 2050 Costa Rica Case Study Johan Rojas, Karen Carranza, and María de los Ángeles Campos. Adv Manuf IV: Vol 5-Biomed Eng: Digit, Sustain Indus Appl 5:3
32. McGee RW, Block WE (2022) On abortion: utilitarianism and deontology. Southern Univ Law Rev 50:10–24. https://ssrn.com/abstract=4545166
33. Lottu OA, Jacks BS, Ajala OA, Okafo ES (2024) Towards a conceptual framework for ethical AI development in IT systems. World J Adv Res Rev 21(3):408–415. https://doi.org/10.30574/wjarr.2024.21.3.0735
34. Sola A (2023) Utilitarianism and consequentialist ethics: framing the greater good. In: Ethics and pandemics: interdisciplinary perspectives on COVID-19 and future pandemics. Springer Nature Switzerland, Cham, pp 61–83
35. Udayakumar L, Babu VSS, Babu VSS (2021) Immanuel Kant's deontology theory. IJRAR-Int J Res Anal Rev (IJRAR) 8(2):235–243
36. Kim B, Wen R, Zhu Q, Williams T, Phillips E (2021) Robots as moral advisors: the effects of deontological, virtue, and confucian role ethics on encouraging honest behavior. In: Companion of the 2021 ACM/IEEE international conference on human-robot interaction (HRI '21 Companion), March 8–11, 2021, Boulder, CO, USA. ACM, New York, NY, USA, p 9
37. Briggle A, Mitcham C (2012) Ethics and science: an introduction. Cambridge University Press. Google-Books-ID: McfZT0AESHwC
38. Kant I (2023) The collected works of Immanuel Kant. Good Press
39. Bradley DA, Seward D, Dawson D, Burge S (2018) Mechatronics and the design of intelligent machines and systems. CRC Press
40. Salvini P, Palmerini E, Koops BJ (2019) Robotics and responsible research and innovation. In: International handbook on responsible innovation. Edward Elgar Publishing, pp 405–424
41. Caram CS, Peter E, Ramos FR, Brito MJ (2022) The process of moral distress development: a virtue ethics perspective. Nurs Ethics 29(2):402–412
42. Yeung K (2018) A study of the implications of advanced digital technologies (including AI systems) for the concept of responsibility within a human rights framework. MSI-AUT 2018:5
43. Mpofu R, Nicolaides A (2019) Frankenstein and the fourth industrial revolution (4IR): ethics and human rights considerations. Afr J Hosp Tourism Leisure 8(5):1–25
44. Floridi L, Cowls J, Beltrametti M, Chatila R, Chazerand P, Dignum V, Vayena E (2018) AI4People—an ethical framework for a good AI society: opportunities, risks, principles, and recommendations. Mind Mach 28:689–707
45. Garrigan B, Adlam AL, Langdon PE (2018) Moral decision-making and moral development: toward an integrative framework. Dev Rev 49:80–100
46. Das SS, Singh K, Kumar R, Singh JN, Kumar P (2024) Exploring the multifaceted landscape of human rights: a comprehensive analysis of theoretical frameworks. Educ Admin Theory Pract 30(5):3346–3365

47. Kent MD, Kopacek P (2021) Social and ethical aspects of automation. In: Digital conversion on the way to Industry 4.0: selected papers from ISPR2020, September 24–26, 2020 Online-Turkey. Springer International Publishing, pp 363–372
48. Chouinard U, Pigosso DC, McAloone TC, Baron L, Achiche S (2019) Potential of circular economy implementation in the mechatronics industry: an exploratory research. J Clean Prod 239:118014
49. Jiménez López E, Cuenca Jiménez F, Luna Sandoval G, Ochoa Estrella FJ, Maciel Monteón MA, Muñoz F, Limón Leyva PA (2022) Technical considerations for the conformation of specific competences in mechatronic engineers in the context of industry 4.0 and 5.0. Processes 10(8):1445
50. Liagkou V, Stylios C, Pappa L, Petunin A (2021) Challenges and opportunities in industry 4.0 for mechatronics, artificial intelligence and cybernetics. Electronics 10(16):2001
51. Hehenberger P, Bradley D (eds) (2016) Mechatronic futures: Challenges and solutions for mechatronic systems and their designers. Springer
52. Balakrishnan S, Lim T, Zhang Z (2022) A methodology for evaluating the economic risks of hurricane-related disruptions to port operations. Transp Res Part A: Policy Pract 162:58–79
53. Collan M, Michelsen KE (2020) Technical, economic and societal effects of manufacturing 4.0: automation, adaption and manufacturing in Finland and beyond. Springer Nature, p 266
54. Bock T (2015) The future of construction automation: Technological disruption and the upcoming ubiquity of robotics. Autom Constr 59:113–121
55. Phan MH, Ngo HQT (2020) A multidisciplinary mechatronics program: From project-based learning to a community-based approach on an open platform. Electronics 9(6):954
56. Holder-Pearson LR (2022) Medical innovation: using mechatronics engineering to reduce inequities in healthcare
57. Luque-Vega LF, Lopez-Neri E, Santoyo A, Ruiz-Duarte J, Farrera-Vazquez N (2019) Educational methodology based on active learning for mechatronics engineering students: towards educational mechatronics. Computación y Sistemas 23(2):325–333
58. Ak A, Topuz V, Altıkardeş A, Oral B (2018) Development of a remote laboratory infrastructure and LMS for mechatronics distance education. EURASIA J Math Sci Technol Educ 14(6):2493–2508
59. Karkoub M, Abdulla S (2020) Transformative learning experiences in mechanical engineering through mechatronics: from high school to college. Int J Mech Eng Educ 48(1):3–31
60. Carlos-Mancilla MA, Luque-Vega LF, Guerrero-Osuna HA, Ornelas-Vargas G, Aguilar-Molina Y, González-Jiménez LE (2020) Educational mechatronics and internet of things: a case study on dynamic systems using Meiot weather station. Sensors 21(1):181
61. Raghunath N, Haapala KR, Sanchez CA (2023) Examining industry expectations for content knowledge in mechatronics across career and professional certificate programs. Manuf Lett 35:1230–1235
62. Cano-Corona A, Hernández-Zempoaltecatl R, Aguilar-Aguilar Á (2021) Working impact in the Puebla-Tlaxcala region of graduates of mechatronics engineering. J Technol Educ 5–14
63. Laužikas M, Miliūtė A, Morozovaitė V, Karpičius D (2021) Effestiveness and efficiency criteria for strategic application of mechatronics in business processes. Insights Reg Develop 3(3):79–105
64. Djuric AM, Jovanovic V, Goris TV (2015) Preparing students for the advanced manufacturing environment through robotics, mechatronics, and automation training
65. Marzano G, Martinovs A, Ušča S (2019) Mechatronics education: needs and challenges. In: Environment. Technologies. Resources. Proceedings of the international scientific and practical conference, vol 2, pp 214–217
66. Andritoiu D, Bazavan LC, Besnea FL, Roibu H, Bizdoaca NG (2018) Agriculture autonomous monitoring and decisional mechatronic system. In: 2018 19th international Carpathian control conference (ICCC). IEEE, pp 241–246
67. Amertet S, Gebresenbet G, Alwan HM, Vladmirovna KO (2023) Assessment of smart mechatronics applications in agriculture: a review. Appl Sci 13(12):7315

68. Kim K, Jung S, Hwang J (2019) Technology convergence capability and firm innovation in the manufacturing sector: an approach based on patent network analysis. R&D Manage 49(4):595–606
69. Albers A, Heimicke J, Trost S (2020) Effects and interactions of agile principles in the process of mechatronic system development: building a basic understanding for adaptive process design. DS 101: Proceedings of NordDesign 2020, Lyngby, Denmark, 12th–14th August 2020, pp 1–12
70. Gumiel JÁ, Mabe J, Jiménez J, Barruetabeña J (2022) Introducing the electronic knowledge framework into the traditional automotive suppliers' industry: from mechanical engineering to mechatronics. Businesses 2(2):273–289
71. Gattulli V, Ottaviano E, Pelliccio A (2018) Mechatronics in the process of cultural heritage and civil infrastructure management. Mechatronics for cultural heritage and civil engineering, pp 1–31
72. Bradley D, Russell D, Ferguson I, Isaacs J, MacLeod A, White R (2015) The internet of things–the future or the end of mechatronics. Mechatronics 27:57–74
73. Parameshwaran R, Baskar C, Karthik T (2015) An integrated framework for mechatronics based product development in a fuzzy environment. Appl Soft Comput 27:376–390
74. Colantoni A, Monarca D, Laurendi V, Villarini M, Gambella F, Cecchini M (2018) Smart machines, remote sensing, precision farming, processes, mechatronic, materials and policies for safety and health aspects. Agriculture 8(4):47
75. Qiu L, Lin X, Wang Y, Yuan S, Shi W (2020) A mechatronic smart skin of flight vehicle structures for impact monitoring of light weight and low-power consumption. Mech Syst Signal Process 144:106829
76. Favi C, Marconi M, Germani M, Mandolini M (2019) A design for disassembly tool oriented to mechatronic product de-manufacturing and recycling. Adv Eng Inform 39:62–79
77. Yang J, Zhang T, Hong J, Zhang H, Zhao Q, Meng Z (2021) Research on driving control strategy and Fuzzy logic optimization of a novel mechatronics-electro-hydraulic power coupling electric vehicle. Energy 233:121221
78. Kamińska D, Zwoliński G, Wiak S, Petkovska L, Cvetkovski G, Barba PD, Anbarjafari G (2021) Virtual reality-based training: case study in mechatronics. Technol Knowl Learn 26:1043–1059
79. Torresen J (2018) A review of future and ethical perspectives of robotics and AI. Front Robot AI 4:75
80. Battistuzzi L, Recchiuto CT, Sgorbissa A (2021) Ethical concerns in rescue robotics: a scoping review. Ethics Inf Technol 23(4):863–875
81. Leenes R, Palmerini E, Koops BJ, Bertolini A, Salvini P, Lucivero F (2017) Regulatory challenges of robotics: some guidelines for addressing legal and ethical issues. Law Innov Technol 9(1):1–44
82. Różańska-Walczuk M (2022) Collaborative robotics. Safety and ethical considerations. In: Machine intelligence and digital interaction conference. Springer Nature Switzerland, Cham, pp 260–269
83. Akpuokwe CU, Adeniyi AO, Bakare SS (2024) Legal challenges of artificial intelligence and robotics: a comprehensive review. Comput Sci IT Res J 5(3):544–561
84. Bissell D, Birtchnell T, Elliott A, Hsu EL (2020) Autonomous automobilities: the social impacts of driverless vehicles. Curr Sociol 68(1):116–134
85. Hörl S, Ciari F, Axhausen KW (2016) Recent perspectives on the impact of autonomous vehicles. Arbeitsberichte Verkehrs-und Raumplanung 1216
86. Schwarting W, Pierson A, Alonso-Mora J, Karaman S, Rus D (2019) Social behavior for autonomous vehicles. Proc Natl Acad Sci 116(50):24972–24978
87. Dubljević V, Douglas S, Milojevich J, Ajmeri N, Bauer WA, List GF, Singh MP (2021) Moral and social ramifications of autonomous vehicles. arXiv preprint arXiv:2101.11775
88. Fagnant DJ, Kockelman K (2015) Preparing a nation for autonomous vehicles: opportunities, barriers and policy recommendations. Transp Res Part A: Policy Practice 77:167–181
89. Riek LD (2017) Healthcare robotics. Commun ACM 60(11):68–78

90. Stahl BC, Coeckelbergh M (2016) Ethics of healthcare robotics: towards responsible research and innovation. Robot Auton Syst 86:152–161
91. Kyrarini M, Lygerakis F, Rajavenkatanarayanan A, Sevastopoulos C, Nambiappan HR, Chaitanya KK, Makedon F (2021) A survey of robots in healthcare. Technologies 9(1):8
92. Kittmann R, Fröhlich T, Schäfer J, Reiser U, Weißhardt F, Haug A (2015) Let me introduce myself: I am Care-O-bot 4, a gentleman robot
93. Guan J (2019) Artificial intelligence in healthcare and medicine: promises, ethical challenges and governance. Chin Med Sci J 34(2):76–83
94. Salem M, Lakatos G, Amirabdollahian F, Dautenhahn K (2015) Towards safe and trustworthy social robots: ethical challenges and practical issues. In: Social robotics: 7th international conference, ICSR 2015, Paris, France, October 26–30, 2015, Proceedings 7. Springer International Publishing, pp 584–593
95. Körtner T (2016) Ethical challenges in the use of social service robots for older adults. Z Gerontol Geriatr 49(4):303–307
96. Wachsmuth I (2018) Robots like me: challenges and ethical issues in aged care. Front Psychol 9:342860
97. Murphy K, Di Ruggiero E, Upshur R, Willison DJ, Malhotra N, Cai JC, Gibson J (2021) Artificial intelligence for good health: a scoping review of the ethics literature. BMC Med Ethics 22:1–17
98. Fletcher SR, Webb P (2017) Industrial robot ethics: the challenges of closer human collaboration in future manufacturing systems. In: A world with robots: international conference on robot ethics: ICRE 2015. Springer International Publishing, pp 159–169
99. Hooker J, Kim TW (2019) Ethical implications of the fourth industrial revolution for business and society. In: Business ethics, vol 3. Emerald Publishing Limited, pp 35–63
100. Acemoglu D, Restrepo P (2020) Robots and jobs: evidence from US labor markets. J Polit Econ 128(6):2188–2244
101. De Vries GJ, Gentile E, Miroudot S, Wacker KM (2020) The rise of robots and the fall of routine jobs. Labour Econ 66:101885
102. Gomes O, Pereira S (2020) On the economic consequences of automation and robotics. J Econ Admin Sci 36(2):135–154
103. Wright SA, Schultz AE (2018) The rising tide of artificial intelligence and business automation: developing an ethical framework. Bus Horiz 61(6):823–832
104. Jaradat O, Sljivo I, Habli I, Hawkins R (2017) Challenges of safety assurance for industry 4.0. In: 2017 13th European dependable computing conference (EDCC). IEEE, pp 103–106
105. Shneiderman B (2020) Human-centered artificial intelligence: reliable, safe & trustworthy. Int J Human-Comput Interact 36(6):495–504
106. Kumar YP, Bhimasingu R (2015) Key aspects of smart grid design for distribution system automation: architecture and responsibilities. Procedia Technol 21:352–359
107. Maddikunta PKR, Pham QV, Prabadevi B, Deepa N, Dev K, Gadekallu TR, Liyanage M (2022) Industry 5.0: a survey on enabling technologies and potential applications. J Indus Inform Integr 26:100257
108. Shneiderman B (2020) Bridging the gap between ethics and practice: guidelines for reliable, safe, and trustworthy human-centered AI systems. ACM Trans Interact Intell Syst (TiiS) 10(4):1–31
109. Auernhammer J (2020) Human-centered AI: the role of Human-centered design research in the development of AI
110. Fokkinga S, Desmet P, Hekkert P (2020) Impact-centered design: introducing an integrated framework of the psychological and behavioral effects of design. Int J Des 14(3):97
111. Melles M, Albayrak A, Goossens, R. (2021). Innovating health care: key characteristics of human-centered design. Int J Qual Health Care 33(Supplement_1):37–44
112. Holeman I, Kane D (2020) Human-centered design for global health equity. Inf Technol Dev 26(3):477–505
113. Lyon AR, Brewer SK, Areán PA (2020) Leveraging human-centered design to implement modern psychological science: return on an early investment. Am Psychol 75(8):1067

114. Kopacek P (2014) Ethical and social aspects of robots. IFAC Proceedings Volumes 47(3):11425–11430
115. Dodig Crnkovic G, Çürüklü B (2012) Robots: ethical by design. Ethics Inf Technol 14:61–71
116. Negi P (2021) Exploring human-robot interaction in mechatronic systems: challenges and opportunities. Math Stat Eng Appl 70(1):386–394
117. Samad T (2023) Human-in-the-loop control and cyber–physical–human systems: applications and categorization. Cyber–physical–human systems: fundamentals and applications, pp 1–23
118. Vanderhaegen F, Jiménez VDB (2023) Opportunities and threats of interactions between humans and cyber–physical systems–integration and inclusion approaches for CPHS. Cyber–physical–human systems: fundamentals and applications, pp 71–90
119. Trentesaux D (2021) Ensuring ethics of cyber-physical and human systems: a guideline. In: Service oriented, Holonic and multi-agent manufacturing systems for industry of the future: proceedings of SOHOMA LATIN AMERICA 2021. Springer International Publishing, pp 223–233
120. Netto M, Spurgeon SK (2017) Special section on cyber-physical & human systems (CPHS). Annu Rev Control 44:249–251
121. Berry C, Gennert M, Reck R (2020) Practical skills for students in mechatronics and robotics education. In: ASEE annual conference exposition proceedings
122. Pannier C, Berry C, Morris M, Zhao X (2020) Diversity and inclusion in mechatronics and robotics engineering education. In: ASEE annual conference exposition proceedings
123. Grimheden ME, Flening E (2022) Developing education in mechatronics to support the challenges for evolution, development, and sustainability. EcoMechatronics: challenges for evolution, development and sustainability. Springer International Publishing, Cham, pp 291–305
124. González-Almaguer CA, Arriaga F, Manriquez J, Acuña A, Gonzalez G, Huerta R, Lloveras J. Teaching ethics in engineering and design, the necessity of concurrent engineering. Des Educ Human Technol Relat 700
125. Jovanovic VM, Michaeli JG, Popescu O, Moustafa MR, Tomovic M, Verma AK, Lin CY (2014) Implementing mechatronics design methodology in mechanical engineering technology senior design projects at the old dominion university
126. Garduño-Aparicio M, Rodríguez-Reséndiz J, Macias-Bobadilla G, Thenozhi S (2017) A multidisciplinary industrial robot approach for teaching mechatronics-related courses. IEEE Trans Educ 61(1):55–62
127. Ponce H, Moya-Albor E, Brieva J (2024) From project-based learning to innovative technologies in mechatronics course: a case study in a private University in Mexico City. In: Data-driven innovation for intelligent technology: perspectives and applications in ICT. Springer Nature Switzerland, Cham, pp 111–134
128. Tutunji TA, Baniyounis M (2018) Measuring quality performance of mechatronics engineering programs. In: 2018 19th international conference on research and education in mechatronics (REM). IEEE, pp 64–68
129. Clò S, Florio M, Pellegrin J, Sirtori E (2018) How do research intensive systems emerge in less developed areas? The case of mechatronics in the Italian southern region Apulia. Int J Technol Learn Innov Develop 10(1):1–36
130. Pop IG, Mătieș V (2011) Transdisciplinary approach of the mechatronics in the knowledge based society. Adv Mechatronics 271–300
131. Pietrusewicz K (2019) Metamodelling for design of mechatronic and cyber-physical systems. Appl Sci 9(3):376
132. Khoshnoud F, Robinson D, De Silva CW, Quadrelli BM (2019) Informed service-learning in mechatronics and dynamic systems. In: 2019 Pacific Southwest section meeting
133. Casale Mashiah D, Beeri I, Vigoda-Gadot E, Hartman A (2023) Responsible research and innovation in Europe: empirical evidence from regional planning initiatives in Austria, Norway, and Spain. Eur Plan Stud 31(9):1949–1974
134. Ollero A, Boverie S, Goodall R, Sasiadek J, Erbe H, Zuehlke D (2006) Mechatronics, robotics and components for automation and control: IFAC milestone report. Annu Rev Control 30(1):41–54

135. Li MM, Datto M, Duncavage EJ, Kulkarni S, Lindeman NI, Roy S, Nikiforova MN (2017) Standards and guidelines for the interpretation and reporting of sequence variants in cancer: a joint consensus recommendation of the Association for Molecular Pathology, American Society of Clinical Oncology, and College of American Pathologists. J Mol Diagn 19(1):4–23
136. Rossi S, Antal A, Bestmann S, Bikson M, Brewer C, Brockmöller J, Hallett M (2021) Safety and recommendations for TMS use in healthy subjects and patient populations, with updates on training, ethical and regulatory issues: expert guidelines. Clin Neurophysiol 132(1):269–306
137. Abbott KW, Snidal D (2021) The governance triangle: Regulatory standards institutions and the shadow of the state. In: The spectrum of international institutions. Routledge, pp 52–91
138. Ruiz V, Pfrang A, Kriston A, Omar N, Van den Bossche P, Boon-Brett L (2018) A review of international abuse testing standards and regulations for lithium ion batteries in electric and hybrid electric vehicles. Renew Sustain Energy Rev 81:1427–1452
139. Gann GD, McDonald T, Walder B, Aronson J, Nelson CR, Jonson J, Dixon K (2019) International principles and standards for the practice of ecological restoration. Restor Ecol 27(S1):S1–S46
140. Swanson KS, Gibson GR, Hutkins R, Reimer RA, Reid G, Verbeke K, Sanders ME (2020) The international scientific association for probiotics and prebiotics (ISAPP) consensus statement on the definition and scope of synbiotics. Nat Rev Gastroenterol Hepatol 17(11):687–701
141. Dwivedi YK, Hughes L, Baabdullah AM, Ribeiro-Navarrete S, Giannakis M, Al-Debei MM, Wamba SF (2022) Metaverse beyond the hype: Multidisciplinary perspectives on emerging challenges, opportunities, and agenda for research, practice and policy. Int J Inf Manage 66:102542
142. Green B (2022) The flaws of policies requiring human oversight of government algorithms. Comput Law Secur Rev 45:105681
143. Meszaros J, Ho CH (2021) AI research and data protection: can the same rules apply for commercial and academic research under the GDPR? Comput Law Secur Rev 41:105532
144. Chico V (2018) The impact of the general data protection regulation on health research. Br Med Bull 128(1):109–118
145. Kolasa K, Mazzi F, Leszczuk-Czubkowska E, Zrubka Z, Péntek M (2021) State of the art in adoption of contact tracing apps and recommendations regarding privacy protection and public health: systematic review. JMIR Mhealth Uhealth 9(6):e23250
146. Vandekerckhove W (2022) Is it freedom? The coming about of the EU directive on whistleblower protection. J Bus Ethics 179(1):1–11
147. Hu L, Chen Y (2018) A short-term intervention for long-term fairness in the labor market. In: Proceedings of the 2018 world wide web conference, pp 1389–1398
148. Magnan M, Menini A, Parbonetti A (2015) Fair value accounting: information or confusion for financial markets? Rev Acc Stud 20:559–591
149. Ozmen E (2023) Sustainability in property taxation: an application for revenue stability and fairness perspectives, the city of Buffalo, NY. Int J Strateg Prop Manag 27(6):391–404
150. Kachelmeier SJ, Limberg ST, Schadewald MS (1991) Fairness in markets: a laboratory investigation. J Econ Psychol 12(3):447–464
151. Ovsyanitskaya, L. Y., & Yurasova, E. V. (2015). Information technologies, mechatronics and robotics as a basis of an interdisciplinary approach to engineering and medical education. *Вестник Южно-Уральского государственного университета. Серия: Образование. Педагогические науки* 7(4):101–106
152. Hou R, Shan N, Hao Z (2015) The construction and evaluation of human resources professional norms model. In: 2015 international conference on mechatronics, electronic, industrial and control engineering (MEIC-15). Atlantis Press, pp 764–767

Case Studies: Application of Mechatronics Engineering

Naheen Ibn Akbar, Protik Barua, Shaekh Mohammad Shithil, Rezwan Us Saleheen, Farhan Mahbub, and Enamul Hoq

Abstract The chapter "Case Studies: Mechatronics Engineering" presented three state-of-the-art mechatronics applications in manufacturing, healthcare, and agriculture. The automated prepreg procedure used in composite manufacturing is demonstrated in the first case study. The second example demonstrates how the combination of robotic arms and computer vision can enhance the accuracy of surgical operations. Lastly, the robotic weeder system in modern farming is covered. These real-world examples show how technologies that combine mechanical, electrical, and computer engineering have improved automation, efficiency, and precision. This chapter aims to highlight the pivotal role of mechatronics engineering in shaping our ever-changing world with an emphasis on the interplay between technology and various sectors.

Keywords Mechatronics · Automated manufacturing · Surgical operation · Robotic weeder system

1 Introduction

In today's fast-changing world of technology, combining mechanical engineering, electronics, computer science, and control systems has created a new field called mechatronics. This interdisciplinary field has revolutionized various sectors, from agriculture to the aerospace industry, by combining traditional mechanical components with advanced electronics and intelligent control systems. Each of the sectors has witnessed significant advancements propelled by innovative mechatronic solutions, leading to enhanced efficiency, precision, and functionality.

In the manufacturing industry, mechatronics has brought a new era of digitalization and automation with the advent of digital twins and collaborative robots. Digital twins are virtual replicas of physical systems, processes, or products that enable real-time simulation, monitoring, and optimization [1]. By creating a digital

N. I. Akbar (✉) · P. Barua · S. M. Shithil · R. U. Saleheen · F. Mahbub · E. Hoq
Department of Mechatronics Engineering, World University of Bangladesh, Dhaka, Bangladesh
e-mail: naheeni.akbar@gmail.com

M. M. Rahman et al. (eds.), *Mechatronics*, Emerging Trends in Mechatronics,
https://doi.org/10.1007/978-981-97-7117-2_9

counterpart of a manufacturing plant or a product, manufacturers can gain valuable insights into performance, predict maintenance needs, and optimize production processes. Additionally, collaborative robots, also known as cobots, have revolutionized manufacturing by facilitating safe and efficient human–robot collaboration [2]. These robots are equipped with advanced sensors and vision systems that enable them to work alongside human operators on the factory floor. Cobots can perform a wide range of tasks, from assembly and packaging to quality inspection, thereby augmenting the capabilities of human workers and improving overall productivity. Furthermore, mechatronics plays a crucial role in the development of smart manufacturing systems, where machines, processes, and data are interconnected to enable autonomous decision-making and adaptive control. Through the integration of sensors, actuators, and control systems, smart manufacturing systems can dynamically respond to changes in production requirements, optimize energy consumption, and ensure consistent product quality.

Mechatronics has also catalyzed significant advancements in medical surgery. CV, for instance, enables surgeons to obtain real-time, high-resolution images of the surgical site, enhancing their ability to perform complex procedures with accuracy and confidence. By overlaying virtual images onto the surgical field, CV systems can also provide surgeons with augmented reality guidance, facilitating precise incisions and sutures. Another groundbreaking application of mechatronics in surgery is continuum robots. Unlike traditional rigid robotic systems, continuum robots are flexible, snake-like devices that can navigate through intricate anatomical structures with remarkable dexterity [3]. These robots are particularly well-suited for minimally invasive procedures, where access to the surgical site is limited. By mimicking the flexibility and adaptability of biological organisms, continuum robots enable surgeons to perform delicate surgeries with minimal trauma to surrounding tissues. Moreover, mechatronic prosthetic devices have transformed the lives of individuals with limb disabilities, restoring mobility, and independence [4]. Modern prostheses incorporate advanced materials, sensors, and actuators to emulate the functionality of natural limbs. By interfacing with the user's nervous system, these prosthetic devices can detect intention signals from the brain and translate them into precise movements, allowing users to perform a wide range of activities with ease.

In agriculture, mechatronics has revolutionized traditional farming practices, paving the way for more efficient, sustainable, and resilient agricultural systems. Farm bots, autonomous robotic systems equipped with sensors, actuators, and navigation capabilities, are at the forefront of this transformation. These robots can autonomously perform a variety of tasks, such as planting, watering, fertilizing, and harvesting crops, thereby reducing the need for manual labor and increasing overall productivity. Also, mechatronics enables the development of robotic systems for weed detection and eradication, reducing the reliance on chemical herbicides and promoting more sustainable farming practices. These robotic weeders can identify and remove weeds with precision, minimizing crop damage and increasing overall efficiency.

The following case studies will focus on three different applications of mechatronics in three distinct fields: manufacturing, medical surgery, and agriculture. We will learn how mechatronic solutions have revolutionized processes, increased precision, and fostered innovation in each domain, illuminating the interdisciplinary field's limitless potential to drive efficiency, efficacy, and sustainability.

2 Case 1: Automated Composite Prepreg Layup

In composite manufacturing, layup process is a method of stacking individual layers or plies of prepregs (fibers pre-impregnated with resin). This process plays a crucial role in producing high-quality composite components for various industries such as aerospace, automotive, and marine. The conventional manual layup process, depicted in Fig. 1a, heavily relies on human labor, where workers manually position each prepreg sheet onto the tooling and apply localized pressure using either their hands or specialized hand tools. However, this method is not only labor intensive but also prone to inconsistencies and defects such as air gaps, wrinkles, and bridging [5]. To address these challenges, researchers have developed an automated system for the robotic layup of composite prepreg sheets [6, 7]. This case study delves into the implementation and benefits of such a system in improving efficiency and consistency in the layup process.

The automated layup process involves a series of steps where robotic systems are utilized to handle and place the prepreg sheets onto a mold or tool in a precise and controlled manner. Here is an overview of how the automated process works:

- **System Initialization**: The automated system is initialized with the layup instructions, including the layup sequence, orientation of each ply (prepreg sheet), and any specific requirements for the composite component being manufactured.

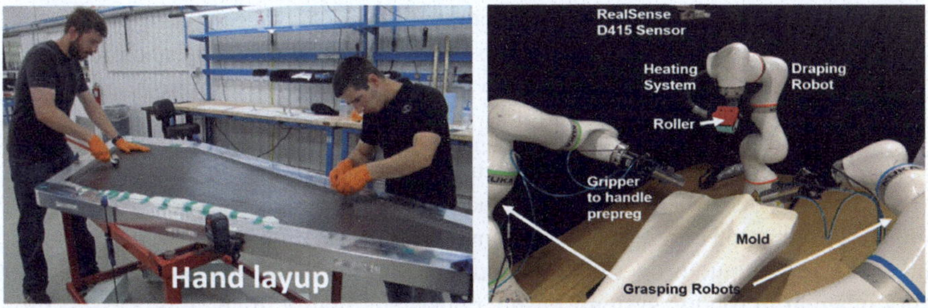

(a) (b)

Fig. 1 **a** Hand layup process [8], **b** Automated layup process with a combination of manipulator, specialized gripper, and conforming tool [7]

- **Material Handling**: Robotic arms equipped with end-effectors designed for handling prepreg sheets pick up individual plies from a staging area or material stack. Vision systems may be used to identify and locate the prepreg sheets accurately.
- **Grasp Planning**: Advanced algorithms are employed to determine the optimal grasp points on the prepreg sheets to ensure secure handling and precise placement. Grasp planning takes into account factors such as material properties, surface conditions, and mold geometry.
- **Compliance Control**: The robotic arms operate in compliance mode, allowing them to adjust their movements and forces in response to external factors such as surface irregularities or material variations. This ensures that the prepreg sheets are placed with the required pressure and alignment.
- **Layer Stacking**: The robotic arms place each prepreg ply onto the mold according to the predefined layup sequence and orientation. The system may incorporate multi-robot collaboration to optimize the layup process and handle complex geometries efficiently.
- **Quality Assurance**: Throughout the layup process, sensors and inspection equipment are used to monitor the placement of the prepreg sheets, detect any defects or deviations, and ensure conformity to the mold surface. This real-time feedback allows for adjustments to be made as needed.
- **Consolidation and Curing**: Once the layup is complete, the composite structure may undergo consolidation processes such as vacuum bagging or autoclave processing to remove air pockets and ensure proper resin flow. The component is then cured under controlled conditions to bond the layers together.
- **Trimming and Finishing**: After curing, the finished composite component is removed from the mold. Any excess material is trimmed, and finishing operations such as sanding or drilling may be performed to achieve the final dimensions and surface quality.
- **Data Logging and Analysis**: The automated system may record data on the layup process, including handling parameters, quality inspection results, and production metrics. This data can be analyzed to optimize the process, improve efficiency, and ensure consistent quality in future manufacturing runs.

The implementation of automated layup system for composite prepreg materials offers a multitude of benefits to manufacturers in various industries. Automation enhances production efficiency by reducing cycle times, minimizing material waste, and enabling continuous operation without interruptions. The precision and repeatability inherent in robotic systems ensure consistent quality in layup processes, ultimately leading to improved product reliability and performance. Moreover, automation significantly reduces labor costs and mitigates ergonomic challenges associated with manual layup, while also lowering the risk of human errors. The automated planning system plays a crucial role in ensuring the consistent placement of composite prepreg sheets, thereby reducing variability and defects in the final components. This consistency is particularly vital for meeting the stringent quality standards prevalent in industries such as aerospace and automotive. Furthermore, the system's flexibility allows for easy reconfiguration to accommodate different mold geometries or composite structures, providing manufacturers with the versatility to produce various components efficiently. Overall, the automated layup system enhances productivity, simplifies manufacturing processes, and facilitates the production of high-quality composite parts with improved accuracy and dependability.

3 Case 2: Surgical Operation Using Robotic Arm and Computer Vision

There is a vast number of scope of Mechatronics in the field of surgical operation like computer vision (CV), haptics, and robotics arm. In contemporary surgical practice, a variety of surgical cameras are employed, offering a rich visual signal that helps surgeons evaluate the clinical site and make clinical judgments. Additionally, this signal can be utilized by artificial intelligence (AI) techniques to assist in real time during procedures and postoperatively for analytics and comprehension of surgical processes by detecting tools, structures, or activities. Advances in CV have made it possible to apply enhanced visualization and augmented reality in the medical field [9]. CV has emerged as a wonderful tool for a wide range of computer-assisted medical applications in recent years. This study uses the example of surgical instrument tracking in microsurgery to illustrate how CV can assist the surgeon during an intervention. Here some examples include requirements and challenges that are common for the use of this technique in a variety of medical applications. Specifically, experts demonstrate how to construct machine learning-based algorithms for simultaneous tool tracking and posture estimation, as well as how to improve resilience in situations involving retinal microsurgery images that have high noise levels and broad fluctuations in light. We also go into detail on how to assess the correctness of an algorithm's overall performance and outline the missing steps required to use these methods in actual clinical settings. A gradual paradigm change in surgery has been toward minimally invasive surgery (MIS) [10]. This indicates that the majority of operating rooms (ORs) have digital cameras installed today, enabling patients to see the operative site. A type of digital measurement and observation of the patient anatomy at the surgical site is produced by the video recorded by the surgical cameras [11]. It provides details about the structure, movement, look, and purpose of the anatomy and equipment inside of it.

When a surgery is documented over the course of the procedure, additional information is embedded regarding the surgical process, activities taken, tools utilized, potential complications or hazards, and even risk assessment. Expert observers may be able to deduce this kind of information, but automated procedures are required to successfully use the data to drive improvements in practice. This is not practicable for normal clinical use. Most surgical video is currently either not recorded at all or is briefly kept on the stack next to the surgical camera before being disposed of at a later time. Surgeons may opt to record their case history on an individual basis, and videos may be utilized for teaching purposes or in case presentations at society conferences or clinical meeting discussions. Since storage is expensive, it makes sense to limit data stores to only include pertinent and medically helpful information. This is mostly because there aren't many tools available that can convert surgical video into interpretable data about the procedure or physiological details seen in the video observations. While it's becoming common practice for some diagnostic procedures—like endoscopic gastroenterology—to store photos from the process into the patient's medical file in order to record lesions identified, surgical video is not usually one of these procedures. Today, it's typical to find additional OR cameras in addition to surgical cameras [10]. These are not limited to the surgical site; they can be used to monitor activity across the entire operating room. Opportunities therefore exist to record this signal and offer a comprehension of the entire space as well as any activities or events that take place there. This might be used to track occurrences at the room level that could help the surgical procedure or to maximize team performance. The development of algorithms for video analysis and comprehension is required in order to utilize video data from the OR in an efficient manner. In order to analyze surgical data, we present a brief overview of the state of the art in AI, with a focus on CV. We also sketch some ideas and future prospects for this field's development and practical application in the clinic [13].

The components which are needed for robotic surgical operation are as follows**:**

- **Robotic Surgical System**: This system encompasses high-tech tools for accurate and minimally invasive procedures. Its core component is the robotic arm illustrated in Fig. 2, which is a multi-jointed mechanical extension equipped with surgical instruments. The surgeon uses a console and associated control interfaces to operate this arm.
- **Cameras**: Real-time photos and videos of the surgical site are taken using high-resolution cameras. These cameras can be positioned to offer various views and angles on the robot's arms or on independent platforms.
- **Image Processing Software**: Real-time photos and videos of the surgical site are taken using high-resolution cameras. These cameras can be positioned to offer various views and angles on the robot's arms or on independent platforms.
- **Depth Sensors**: Depth information about the surgical scene is provided by depth sensors, such as time-of-flight cameras or structured light systems. For applications such as collision avoidance, depth perception, and 3D reconstruction, this depth data is essential.
- **Stereo Vision Systems**: The cameras in a stereo vision system are arranged in pairs to simulate human binocular vision. Stereo vision systems evaluate depth information and generate three-dimensional (3D) representations of the surgical environment by triangulating the differences between corresponding spots in the pictures from each camera.
- **Machine Learning Models**: Deep learning algorithms and other machine learning approaches are used for tasks including segmentation, prediction, classification, and object detection. These models are trained using data and annotated

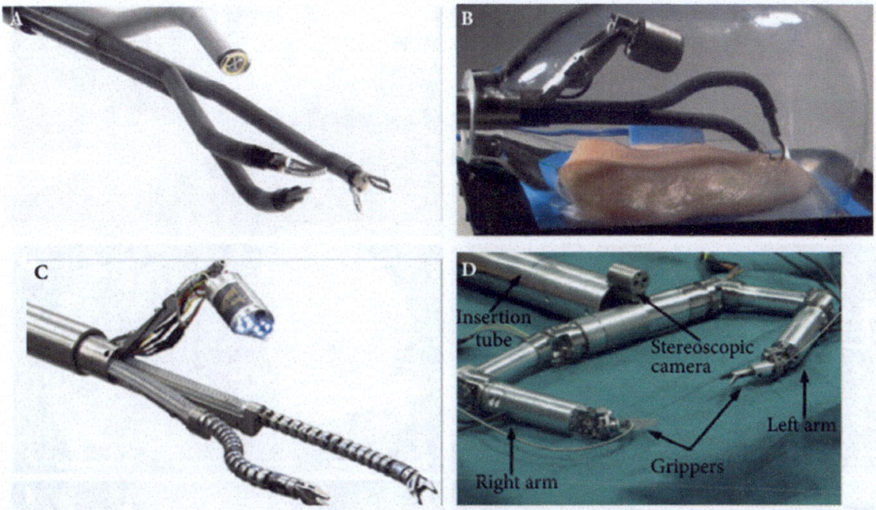

Fig. 2 Different types of end-effectors for robotic surgical operation [12]

surgical images to identify trends and make wise choices while the surgery is being performed.

- **Augmented Reality (AR) Displays**: AR displays superimpose visual cues, data, or computer-generated pictures over the surgeon's actual view of the operating scene. To help the surgeon during the procedure, this technology offers real-time assistance, anatomical information, and procedural instructions.
- **Tracking Systems**: Real-time tracking systems keep an eye on the position and orientation of the robot, anatomical structures, and surgical equipment. This data is essential for ensuring precise spatial registration throughout the process and lining up virtual overlays with the real world.
- **User Interfaces**: Surgeons and other medical professionals can see the data that has been gathered, control the movements of the robot, and communicate with the CV system using user interfaces. In the surgical setting, intuitive interfaces that include voice commands, gesture detection, haptic feedback, and touchscreens improve workflow efficiency and user friendliness (Fig. 3).

The overall process for surgical operation is given below sequentially:

- **Image Acquisition**: Real-time photos and videos of the surgical site are taken by high-resolution cameras that are either placed throughout the operating room or mounted on the surgical robot. Infrared, visible light, and other specialist imaging modalities could be included in this group of cameras [11].
- **Image Preprocessing**: The cameras' raw pictures undergo preprocessing in order to improve their quality and get them ready for additional examination. Preprocessing could include things like color correction, image registration, contrast improvement, and noise reduction.

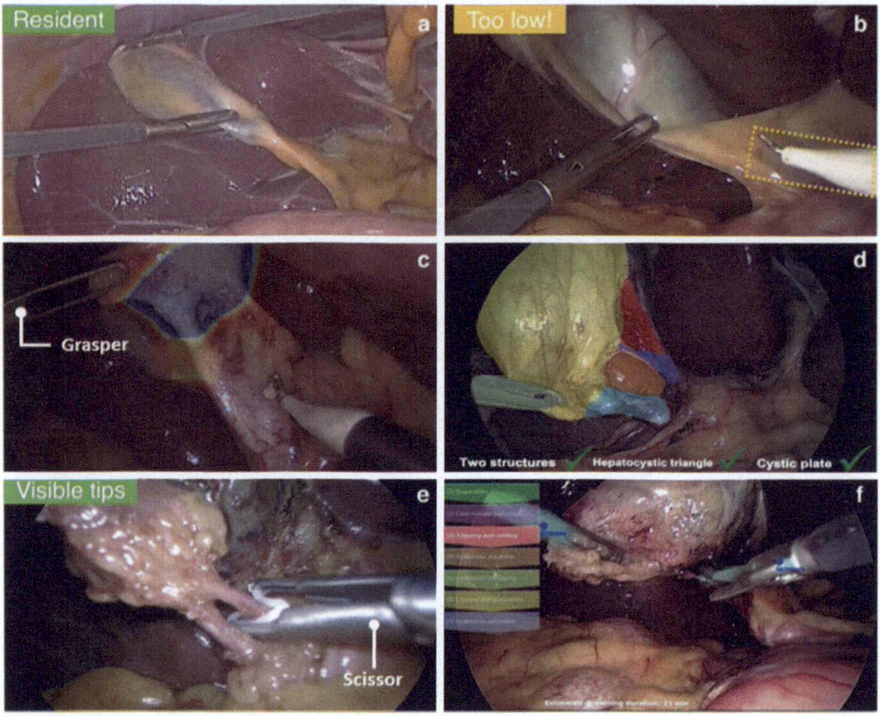

Fig. 3 CV-based surgical operation [14]

- **Feature Extraction**: Preprocessed images are analyzed by CV algorithms to extract pertinent features including textures, forms, edges, and corners. These characteristics aid in the identification and characterization of various anatomical components, surgical tools, and other surgical scene objects.
- **Object Detection and Segmentation**: Sophisticated algorithms identify and separate particular things of interest, including organs, blood arteries, malignancies, or surgical equipment, from the operative images. To precisely identify and label individual items, methods such as object detection, semantic segmentation, and instance segmentation are used.
- **3D Reconstruction**: A three-dimensional (3D) model of the surgical environment is created using depth data from depth sensors or stereo vision equipment. Spatial context and depth perception are provided by this 3D reconstruction, which improves the robot's ability to navigate and engage with the environment.
- **Object Tracking and Motion Estimation**: Real-time tracking of the position and motion of dynamic objects such as surgical instruments and anatomical structures is facilitated by CV algorithms. The robot can stay aware of its surroundings, predict movements, and modify its behavior as a result of this tracking.
- **Image Analysis and Decision-Making**: Machine learning models make smart decisions and support the surgeon during the process by analyzing the segmented

objects, reconstructed 3D models, and extracted data. These models can be used for applications including surgical gesture recognition, instrument tracking, tissue classification, and pathology identification.

- **Augmented Reality (AR) Visualization**: The surgeon's vision of the actual surgical scene is enhanced with computer-generated visuals, data, or visual cues through the use of augmented reality. AR displays help improve surgical precision and decision-making by offering real-time advice, anatomical information, and procedural instructions.
- **Integration with Surgical Workflow**: Surgeons and medical staff can interact with the visual data, manipulate the movements of the robot, and retrieve pertinent information during an operation with ease thanks to CV systems' seamless integration into the surgical workflow.

Overall, CV offers a novel approach to researching and improving the intraoperative phase of surgical operation with robotics arm. This operation is comparatively safe and comfortable for patients. As the clinical and data science communities have begun to coalesce around enhancing research and scientific inquiry on the optimal application of CV in surgery, numerous proof-of-concept applications with potential therapeutic benefits have been produced in MIS. Efforts to generalize such applications have primarily focused on enhancing modeling methodologies and streamlining access to surgical data, while always considering the ethical and cultural aspects that are critical to patient care. As CV in surgery advances and complicated operation, more social and clinical engagement will be needed to ensure that the field's potential is fulfilled in a safe and efficient manner, supporting the treatment of surgical patients.

4 Case 3: Automated Weed Removal System in Agriculture

The large-scale agricultural farm usually covered hundreds of acres and faced trouble with lots of weeds growing, which hurt how many crops it could grow and how productive it was overall. Using traditional ways like pulling weeds by hand or spraying lots of pesticides everywhere took a lot of work, cost a lot of money, and was not environmentally sustainable. To address these challenges, the farm adopted an automated weed removal system comprising autonomous mobile robots equipped with advanced sensors and AI algorithms shown in Fig. 4.

These robots were designed to navigate the vast farmland, identify weeds with precision, and execute targeted removal strategies. With this new technology, farmers can handle weed problems better without needing to do as much hard work themselves. It's also better for the environment. This case study underscores the potential of automated weed removal solutions to revolutionize weed control practices in the modern agricultural landscape.

The key components consisting of an automated weeder system are the following:

(a) (b)

Fig. 4 Weeding robots **a** EcoRobotix [15] and **b** Thorvald II [16]

- **Robotics Platform**: The autonomous robots, guided by GPS and onboard sensors, traversed the fields systematically, scanning for weed presence.
- **CV**: Utilizing sophisticated CV algorithms, the robots analyzed real-time images to differentiate between crops and weeds accurately.
- **Machine Learning Model**: By leveraging machine learning models trained on diverse weed species and environmental conditions, the system continuously improved its weed detection capabilities.
- **Weed Removal Mechanism**: Upon identifying weeds, the robots deployed precise removal methods, such as mechanical weeding tools or targeted herbicide application, minimizing collateral damage to crops.

The operational workflow displayed in Fig. 5 seamlessly integrates autonomous robots equipped with advanced sensors and AI algorithms to scan fields, identify weeds through real-time image analysis, and make informed decisions on targeted weed removal strategies. This makes managing weeds in agriculture much more efficient and precise. Additionally, this method brings several advantages, which are discussed below.

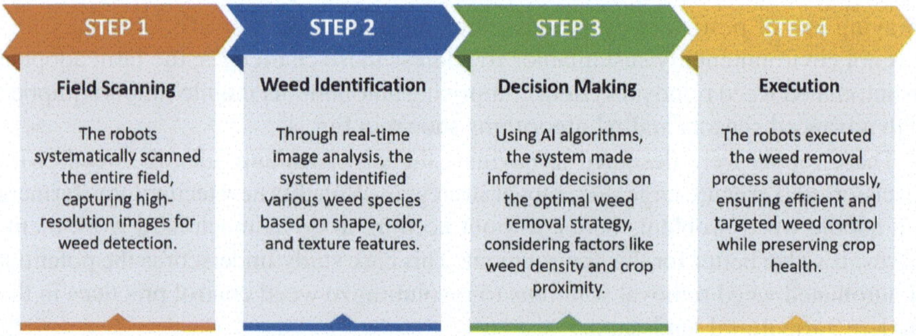

Fig. 5 Operational workflow of automatic weeding process

- **Enhanced Operational Efficiency**: The implementation of automated weed removal processes significantly enhances operational efficiency by reducing the time and labor traditionally required for manual weeding tasks. Autonomous robots can cover large areas efficiently, leading to streamlined weed management operations.
- **Cost Savings**: By minimizing the reliance on manual labor and optimizing resource utilization, automated weed removal systems offer substantial cost savings to farmers. The reduction in labor costs and the efficient use of herbicides contribute to overall cost-effectiveness in weed management practices.
- **Environmental Sustainability**: Automated weed removal processes promote environmental sustainability by reducing the need for widespread herbicide application. Targeted weed removal strategies minimize the environmental impact of chemical usage, preserving soil health and biodiversity in agricultural ecosystems.
- **Scalability and Adaptability**: Automated weed removal systems are scalable and adaptable to varying field conditions and crop types. The flexibility of these systems allows farmers to customize weed management strategies based on specific requirements, ensuring optimal results across diverse agricultural settings.
- **Labor Safety and Health Benefits**: By reducing the need for manual labor in weed removal tasks, automated systems contribute to improved labor safety and health outcomes for farm workers. It minimizes exposure to potentially harmful chemicals and repetitive manual labor tasks and enhances the overall well-being of agricultural workers.

As a whole, weeding with the help of mechatronic systems signifies a paradigm shift in agricultural practices. Farmers can achieve sustainable crop production, reduce environmental risks, and meet the growing demands of an ever-changing agricultural landscape by incorporating intelligent weed management systems.

References

1. Jiang, Yin Y, Li S, Luo K, Kaynak H, O (2021) Industrial applications of digital twins philosophical transactions of the royal society: A Mathematical. Phys Eng Sci 379(2207):20200360. https://doi.org/10.1098/rsta.2020.0360
2. Pauliková A, Gyurák Babeľová, Z, Ubárová, M (2021). Analysis of the impact of human–cobot collaborative manufacturing implementation on the occupational health and safety and the quality requirements. Int J Environ Res Public Health 18(4):4. https://doi.org/10.3390/ijerph18041927
3. Dupont PE, Simaan N, Choset H, Rucker C (2022) Continuum robots for medical interventions. Proc IEEE 110(7):847–870. https://doi.org/10.1109/JPROC.2022.3141338
4. Desplenter T, Zhou Y, Edmonds BP, Lidka M, Goldman A, Trejos AL (2020) Rehabilitative and assistive wearable mechatronic upper-limb devices: A review. J Rehabil Assist Technol Eng 7:2055668320917870. https://doi.org/10.1177/2055668320917870
5. Chen Y-W, Joseph RJ, Kanyuck A, Khan S, Malhan RK, Manyar OM, McNulty Z, Wang B, Barbič J, Gupta SK (2021) A digital twin for automated layup of prepreg composite sheets. J Manuf Sci Eng 144(041010). https://doi.org/10.1115/1.4052132

6. Ellekilde L-P, Wilm J, Nielsen OW, Krogh C, Kristiansen E, Gunnarsson GG, Stenvang TS, Jakobsen J, Kristiansen M, Glud JA, Hannemose M, Aanæs H, de Kruijk J, Sveidahl I, Ikram A, Petersen HG (2021) Design of automated robotic system for draping prepreg composite fabrics. Robotica 39(1):72–87. https://doi.org/10.1017/S0263574720000193
7. Malhan RK, Shembekar AV, Joseph RJ, Kabir AM, Bhatt PM, Shah B, Zhao W, Nutt S, Gupta SK (2020) A smart robotic cell for automating composite prepreg layup—SAMPE. https://www.nasampe.org/store/viewproduct.aspx?id=16317618
8. Modernmachineshop (2018) Managing risk is this shop's bold strategy. https://www.mmsonline.com/articles/managing-risk-is-this-shops-bold-strategy
9. Weiser TG, Regenbogen SE, Thompson KD, Haynes AB, Lipsitz SR, Berry WR, Gawande AA (2008) An estimation of the global volume of surgery: A modelling strategy based on available data. Lancet (London, England) 372(9633):139–144. https://doi.org/10.1016/S0140-6736(08)60878-8
10. Bartoli A, Collins T, Bourdel N, Canis M (2012) Computer assisted minimally invasive surgery: Is medical computer vision the answer to improving laparosurgery? Med Hypotheses 79(6):858–863. https://doi.org/10.1016/j.mehy.2012.09.007
11. Kennedy-Metz LR, Mascagni P, Torralba A, Dias RD, Perona P, Shah JA, Padoy N, Zenati MA (2021) Computer vision in the operating room: Opportunities and caveats. IEEE Trans Med Robot Bionics 3(1):2–10. https://doi.org/10.1109/tmrb.2020.3040002
12. Rassweiler JJ, Autorino R, Klein J, Mottrie A, Goezen AS, Stolzenburg J-U, Rha KH, Schurr M, Kaouk J, Patel V, Dasgupta P, Liatsikos E (2017) Future of robotic surgery in urology. BJU Int 120(6):822–841. https://doi.org/10.1111/bju.13851
13. Chadebecq F, Vasconcelos F, Mazomenos E, Stoyanov D (2020) Computer vision in the surgical operating room. Visc Med 36(6):456–462. https://doi.org/10.1159/000511934
14. Mascagni P, Alapatt D, Sestini L, Altieri MS, Madani A, Watanabe Y, Alseidi A, Redan JA, Alfieri S, Costamagna G, Boškoski I, Padoy N, Hashimoto DA (2022) Computer vision in surgery: From potential to clinical value. Npj Digital Medicine 5(1):1–9. https://doi.org/10.1038/s41746-022-00707-5
15. Ben-Ari M, Mondada F (2018) Robots and their applications. In: Ben-Ari M, Mondada F (Eds.) Elements of Robotics (pp. 1–20). Springer International Publishing. https://doi.org/10.1007/978-3-319-62533-1_1
16. Grimstad L, From PJ (2017) The thorvald II agricultural robotic system. Robotics 6(4):4. https://doi.org/10.3390/robotics6040024

Conclusion and Future Trends of Mechatronics

Mustafizur Rahman, Md. Mizanur Rahman, M. R. Sohel, Ahmed Farhan, Enamul Huq, and Farhan Mahbub

1 Conclusion

Mechatronics is one of the most promising and potentially useful topics in engineering. The intricate network of specialties within the fascinating field has been thoroughly explored across multiple chapters. Mechatronics is the synthesis of several sciences, such as electronic and mechanical engineering, computer science, and control systems, with the goal of building intelligent, adaptive systems that can transform industries and improve our daily lives. It's clear from our analysis of potential developments in mechatronics engineering that this field is headed for explosive growth and innovation.

The most notable trend influencing mechatronics in the future is the speed at which artificial intelligence (AI) and machine learning are developing. Mechatronic systems are progressively being equipped with AI algorithms, which allow them to perceive and react to complicated situations with previously unheard-of accuracy and efficiency. Artificial Intelligence (AI) is enabling robots with cognitive capacities that were previously limited to humans, opening up new possibilities in mechatronics. Examples of this include driverless cars navigating congested city streets and robotic assistants performing delicate surgical procedures.

Moreover, the convergence of edge computing and the Internet of Things (IoT) with mechatronics is opening up exciting new possibilities for networked, intelligent systems. Mechatronic systems can gather enormous volumes of data in real time with the use of IoT sensors and actuators, facilitating improved user experiences, predictive maintenance, and energy consumption optimization. By processing data at the source and decreasing latency and reliance on centralized infrastructure, edge computing's spread further improves the scalability and responsiveness of mechatronic systems.

M. Rahman (✉) · Md. M. Rahman · M. R. Sohel · A. Farhan · E. Huq · F. Mahbub
Department of Mechatronics Engineering, World University of Bangladesh, Dhaka, Bangladesh
e-mail: mustafizur.rahman@mte.wub.edu.bd

M. M. Rahman et al. (eds.), *Mechatronics*, Emerging Trends in Mechatronics,
https://doi.org/10.1007/978-981-97-7117-2_10

Simultaneously, the discipline of robotics keeps expanding the frontiers of mechatronics. Technological developments in material science, sensing, and bioinspired design are opening up new possibilities for the creation of robots that can function in a variety of settings, from the surface of far-off planets to the ocean's depths. Collaborative robots, also known as Cobots, are collaborating with human operators to increase efficiency and safety in industrial settings, thereby transforming manufacturing. Robots are being used more and more in industries like healthcare, agriculture, and disaster relief because they can supplement human abilities and reduce risks. This is because robots are becoming more agile and versatile.

Another important development in mechatronics engineering is the focus on environmental stewardship and sustainability. There is an increasing need for mechatronic solutions that are recyclable, low-impact on the environment, and energy-efficient as the globe struggles with resource depletion and climate change. Engineers are making mechatronic systems both technologically innovative and environmentally responsible by integrating eco-design and circular economy ideas from the design phase to end-of-life disposal.

Furthermore, the democratization of mechatronics is promoting an innovative and entrepreneurial culture, with DIY enthusiasts and companies spearheading community-driven innovation, open-source hardware, and grassroots prototype projects. With the help of platforms like Arduino and Raspberry Pi, mechatronic experimentation has become more accessible, enabling people with a variety of backgrounds to work together on projects ranging from wearable technology to home automation.

Nevertheless, it's important to recognize the moral and societal ramifications of mechatronics engineering in the middle of these fascinating advances. Concerns about employment displacement, security, and privacy are major issues as intelligent robots grow more and more ingrained in our daily lives. To guarantee that mechatronic systems are developed and implemented in a way that puts human well-being, equity, and social responsibility first, engineers and legislators must collaborate.

To sum up, mechatronics engineering has a promising future ahead of it. The field of mechatronics is changing at an astounding rate, from AI-driven robots to networked IoT ecosystems, from sustainable design principles to grassroots creativity. By keeping up with new developments, encouraging cross-disciplinary cooperation, and adhering to moral principles, we may harness the transformative potential of mechatronics to build a more prosperous, just, and sustainable future for future generations. Remember that the real test of success is not just how far technology has come, but also how much of a positive influence we have on the world and the lives of those around us, as we set out on this path of discovery and invention.

2 Future Trends in Mechatronics Engineering

Understanding and strengthening engineering methodologies for systems is becoming more and more dependent on the current trend toward more complicated and diverse system development. The efficient and fruitful administration of these systems will follow from this. Both academics and businesses face new problems as a result of scientific breakthroughs in areas such as computer engineering, simulation and modeling, power electronics, microprocessors, electromechanical motion tools, micro-electro-mechanical systems (MEMS), and distributed system platforms (DSPs). Advanced mechatronic product design relies heavily on modeling, simulation, analysis, visualization, and virtual prototyping. The use of new technology to create smarter, more scalable, multifunctional items that are better, cheaper, and more widely used is a key component of global market competition. This type of system development requires methodologies that are both flexible and adaptable due to the wide range of possible applications in fields such as automotive, aeronautics, robotics, consumer goods, and many more. A mechatronic system is one that combines software with electronic and mechanical components in a way that is both dynamic and multidisciplinary. A variety of development and integration strategies are emerging from the relevant disciplines in an effort to tackle the inherent complexity of the field's elements.

An approach to mechatronics design at the system level is necessary due to the exponential growth in device and system complexity in recent years. Engineers can streamline test execution, reuse algorithms in the final embedded delivery framework, and integrate mechanical and control design with this method. Innovations funding for healthcare, the life sciences, and renewable energy sources, along with innovations in manufacturing gear, are driving this trend at the system level. This method streamlines design, prototyping, and implementation by integrating best practices and existing technologies, which substantially improves the design process. A more streamlined method of production might be implemented by the engineers by dividing the design process into parallel threads. The development of a control algorithm for a mechanical device is used to require the team to wait for a working prototype. Engineers can now begin work more quickly by using a virtual prototype built from concept models and data from simulations. Figure 1 displays the fundamental idea of mechatronics along with present and future mechatronics sectors that have emerged due to technological advancements. Despite the fact that many various visions of the future have been put forth regarding technology and its impact on society, funding is allocated to futurists and tech businesses with the express purpose of predicting the world's shape in the next ten or fifteen years.

Fig. 1 Emerging themes in mechatronics based on fundamental interdisciplinary synergy

2.1 Future Trend in Mechatronics Engineering on Fourth Industrial Revolution 4.0

Mechatronics engineering was leading the way in innovation during the Fourth Industrial Revolution, which involves the integration of digital, physical, and biological technologies. There are several significant trends in mechatronics engineering that are influencing the future of manufacturing, automation, and technology-driven industries.

2.1.1 Advanced Automation and Robotics

Engineers in the field of mechatronics are creating advanced automation and robotic systems to enhance efficiency, flexibility, and productivity in manufacturing operations. Utilizing technologies like artificial intelligence, machine learning, and Internet of Things (IoT) connectivity, these systems enable autonomous operation, predictive maintenance, and real-time optimization.

2.1.2 Digital Twin Technology

Virtual representations of physical assets are transforming product design, prototyping, and maintenance processes. Engineers in the field are using digital twin technology to simulate and optimize the performance of mechatronic systems, predict failures, and enable remote monitoring and control. This results in quicker time-to-market, decreased expenses, and enhanced product quality.

2.1.3 Cyber-Physical Systems (CPS)

Engineers in the field of mechatronics are currently working on developing Cyber-Physical Systems (CPS) for various uses, such as smart factories, self-driving cars, and intelligent infrastructure. These systems facilitate smooth communication between the digital and physical realms, resulting in enhanced efficiency, safety, and dependability.

2.1.4 Additive Manufacturing (3D Printing)

The revolutionary technologies in additive manufacturing are reshaping conventional manufacturing methods, allowing for quick prototyping, personalized products, and just-in-time production. Engineers in the field of mechatronics are currently working on cutting-edge 3D printing technology that can produce intricate designs, multi-material parts, and functional prototypes with exceptional accuracy and speed. This promotes innovation, minimizes waste, and decreases production expenses.

2.1.5 Human–Machine Interaction (HMI)

With the increasing presence of automation and robotics, the relationship between humans and machines is changing. Mechatronics engineers are developing user-friendly and adaptable human–machine interfaces (HMIs) to improve user experience, promote teamwork, and guarantee safety in human–robot interaction. These interfaces encompass touchscreens, gesture recognition, voice commands, and augmented reality displays.

Summarizing the upcoming direction of mechatronics engineering in the Fourth Industrial Revolution, it is marked by cutting-edge automation and robotics, digital twin technology, cyber-physical systems, additive manufacturing, and human–machine interaction. As these trends continue to evolve, individuals in the field will play a pivotal role in driving innovation, transforming industries, and shaping the future of technology-driven societies.

2.2 Future Trend in Mechatronics Engineering on Robotic Revolution

As the robotic revolution continues, mechatronics engineering—an interdisciplinary discipline that brings together mechanical, electrical, and computer engineering—is set to become increasingly important. A number of important tendencies are developing in mechatronics engineering, which are influencing the trajectory of robotics in the future, thanks to the exponential growth of technology.

2.2.1 Integration of Artificial Intelligence (AI)

With the help of AI, robots can now sense their surroundings, make decisions, and adjust their actions appropriately, completely changing the robotics industry. Machine learning, deep learning, and other artificial intelligence algorithms are finding more and more applications in robotic systems designed by mechatronics engineers. These algorithms allow robots to learn from their experiences, make better decisions, and increase their autonomy.

2.2.2 Collaborative Robotics (Cobots)

Big, pricey, and sometimes housed in protective enclosures away from human employees, traditional industrial robots have their limitations. But cobots—robots that can safely and effectively operate alongside humans—are in high demand. With the help of cutting-edge sensing, control, and safety systems, mechatronics engineers are creating robots that can work closely with humans in a wide range of service and industrial contexts.

2.2.3 Miniaturization and Modularization

Robots that are both smaller and nimbler are becoming a reality thanks to developments in miniaturization technology. Mechatronics engineers are taking advantage of miniaturization to create robots that can operate in confined spaces, which are finding more and more uses in fields including medical surgery, infrastructure inspection, and exploration of dangerous or inaccessible areas. The ability to quickly prototype and tailor robotic systems to particular activities and settings is also being made possible by modular design approaches.

2.2.4 Sensory Fusion and Perception

Cameras, light detection and ranging (LiDAR), radar, and inertial measurement units are just a few of the sensors that robots are being outfitted with to better understand and navigate their environments. To help robots independently navigate complicated and ever-changing environments, mechatronics engineers are creating perception and sensor fusion algorithms. These algorithms will allow robots to correctly interpret sensor input, create maps of their surroundings, and more.

2.2.5 Humanoid and Biomimetic Robotics

Biomimetic and humanoid robots take design cues from the natural world and attempt to simulate its structure and behavior. In their quest to create intelligent, agile, and dexterous robots, mechatronics experts are investigating novel materials, actuators, and control systems. Search and rescue missions in the event of a disaster and assistive robotics for the handicapped and old are two examples of the many uses for these robots.

2.2.6 Soft Robotics and Swarm Robotics

In the area of mechatronics, soft robotics and swarm robotics are cutting-edge subfields that offer new capabilities and uses. Soft robotics is the field that studies making robots out of flexible materials like rubber and silicone so that they can do delicate jobs and move around in complicated environments. This technology looks like it could be very useful for medical procedures, assistive devices, and automating the handling of fragile items in factories. Swarm robotics is based on the way social insects act as a whole. It includes controlling many simple robots at once so that they can do complicated tasks more efficiently and reliably. Monitoring the climate, farming, and logistics are some examples of uses. When these technologies come together, they might create robotic systems that can adapt and work together to solve problems. This would make big steps forward in many fields and change the future of mechatronics.

In a nutshell, development in humanoid and biomimetic robots, improvements in sensory fusion and perception, miniaturization, collaborative robotics, and AI integration will define the future trend of mechatronics engineering within the robotic revolution. The future of robotic systems that revolutionize industries, boost human productivity, and improve quality of life is being shaped by mechatronics engineers, who will be essential as these trends progress.

2.3 Future Trend in Mechatronics Engineering on Energy Efficiency and Sustainability

Integrating renewable energy sources, creating energy-efficient designs, and developing sophisticated control systems are the future directions of energy efficiency and sustainability in mechatronics engineering. Smart systems that integrate renewable energy sources like wind and solar power and use sensors and actuators to optimize energy use in real time are being developed by engineers. Energy harvesting systems, which collect and transform waste energy into usable electrical power, are also receiving more and more attention. Engineers attempt to minimize environmental impact from design to disposal by examining the whole lifecycle of systems and utilizing lifecycle assessment approaches. This paves the way for more environmentally friendly and sustainable mechatronic systems.

2.3.1 Smart Energy Management Systems

In order to maximize the efficiency with which buildings of all types and sizes use their energy resources, mechatronics engineers are creating smart energy management systems. By combining sensors, actuators, and control algorithms, these systems can track energy usage, detect inefficiencies, and optimize settings automatically. These systems minimize waste and carbon emissions by maximizing energy efficiency.

2.3.2 Renewable Energy Integration

Renewable energy sources, including solar, wind, and hydroelectric power, rely heavily on mechatronics engineering for integration and control. In order to make the most of renewable energy sources and lessen the impact of their intermittency, mechatronics engineers create and improve power electronics, energy storage systems, and grid integration technologies. A decentralized energy infrastructure that is both more sustainable and easier to implement is made possible by this.

2.3.3 Energy-Efficient Transportation

Electric vehicles (EVs), hybrid vehicles, and driverless vehicles are just a few examples of the eco-friendly and energy-efficient transportation systems that mechatronics experts are creating. These vehicles are more efficient, use less fuel, and produce less emissions because they use sophisticated mechatronic technologies for propulsion, energy storage, and control. Intelligent traffic control systems and vehicle-to-grid (V2G) communication technologies are only two examples of the smart transportation infrastructure that mechatronics engineering helps build.

2.3.4 Energy Harvesting and Waste Utilization

Mechatronics researchers are looking into new ways to harness energy from things like light, heat, and vibrations in the environment. Wireless networks, wearable gadgets, and sensors that can power themselves are now within reach, thanks to advancements in energy harvesting and efficient energy storage. An further way mechatronics engineering helps with sustainability and resource conservation is by making it easier to use renewable energy sources and waste heat from industrial processes.

2.3.5 Green Manufacturing and Process Optimization

Green manufacturing techniques and energy efficiency optimization in industrial processes are greatly assisted by mechatronics engineering. In order to maximize efficiency, decrease waste, and lessen the impact on the environment, mechatronics engineers create smart control systems, predictive maintenance algorithms, and actuators that use less energy. Sustainable manufacturing technologies, like modular production systems and additive manufacturing, are also designed and optimized with the help of mechatronics engineering.

In conclusion, green manufacturing methods, energy harvesting and waste utilization, energy-efficient mobility, smart energy management systems, and renewable energy integration are all parts of the energy efficiency and sustainability trend in mechatronics engineering. A more sustainable and resilient future is being ushered in by mechatronics engineers who are embracing these trends and inventing novel solutions.

2.4 Future Trend of Mechatronics Engineering on Precision Medicine and Medical Devices/Health Industry

When it comes to developing cutting-edge medical gadgets and precision medicine, mechatronics engineering is where it's at. The future of healthcare delivery and individualized treatment is being shaped by several important developments that are emerging in this field.

2.4.1 Miniaturization and Wearable Health Technologies

Wearable health monitoring devices are being developed by mechatronics engineers who are making use of developments in downsizing and sensor technology. These gadgets will give users real-time data on their vital signs, physical activity, and other health factors. By allowing for early identification of health problems, specific therapeutic tactics, and continuous remote monitoring, these gadgets pave the way for precision medicine approaches that are customized to each patient.

2.4.2 Robot-Assisted Surgery and Interventional Procedures

Robotic surgical systems and interventional devices, which allow less invasive treatments with more accuracy and better patient outcomes, are developed in large part by mechatronics engineers. Surgeons are given the ability to make better decisions in real time and have their surgical tools become better and more accurate thanks to these systems that combine imaging technology with sophisticated sensors and actuators.

2.4.3 Telemedicine and Remote Healthcare Delivery

Virtual consultations, remote monitoring, and tele-operated treatments are all within the realm of possibility thanks to the work of mechatronics engineers who are creating telemedicine platforms and other remote healthcare technology. These innovations help people in underserved or faraway places get the healthcare they need by using immersive interfaces, haptic feedback systems, and high-speed communication networks.

2.4.4 Implantable and Prosthetic Devices

Advanced prosthetic limbs and orthotic devices, as well as implantable medical devices like neuro-stimulators, medication delivery systems, and pacemakers, are all products of mechatronics engineering. To improve functionality, biocompatibility, and patient comfort, these devices use smart materials, bio-sensing capabilities, and closed-loop control algorithms. As a result, people with disabilities are able to live more independently and actively.

2.4.5 Personalized Drug Delivery and Therapy

In order to optimize treatment regimens according to specific patient traits and illness profiles, mechatronics engineers are investigating new drug delivery technologies and individualized therapeutic solutions that permit targeted drug delivery, controlled release kinetics, and tailored medication schedules. In order to optimize treatment efficacy while reducing adverse effects, these technologies can adapt to changing physiological conditions in real time. Examples of such devices include smart drug delivery systems, implanted pumps, and microfluidic devices.

Wearable health technologies, telemedicine platforms, implanted devices, prosthetics, robot-assisted surgery, and individualized medication delivery systems are all part of the future trend in mechatronics engineering for precision medicine and medical equipment. Innovation toward more personalized, precise, and patient-centric healthcare solutions is being driven by mechatronics engineers, who integrate multidisciplinary knowledge in engineering, biology, and medicine.

2.5 Future Trend of Mechatronics Engineering on Space Exploration and Aerospace

Spacecraft design, autonomous navigation, and robotics for use on other planets are all areas where mechatronics engineering is paving the way for promising future developments in aerospace technology and space travel. The area of mechatronics engineering is being influenced by a number of important trends.

2.5.1 Autonomous Spacecraft and Rovers

To explore and maneuver heavenly entities like planets, moons, and asteroids, mechatronics engineers are building autonomous spacecraft and rovers with sophisticated sensors, actuators, and control systems. Spacecraft and rovers are able to navigate complex and ever-changing settings with ease thanks to these autonomous systems, which allow for real-time decision-making, adaptive planning, and self-correction.

2.5.2 Additive Manufacturing and In-Situ Resource Utilization

In order to fabricate spacecraft components and infrastructure on-site using locally accessible materials, mechatronics engineers have developed additive manufacturing technologies and in-situ resource utilization (ISRU) systems. Rapid prototyping, modification, and repair of spacecraft parts are made possible by additive manufacturing techniques like 3D printing. On the other hand, ISRU technologies allow for the extraction and exploitation of resources like water, minerals, and gases from planetary surfaces, which are necessary for long-duration missions.

2.5.3 Space Robotics and Manipulation Systems

For jobs like servicing satellites, assembling massive structures, and collecting samples from planetary surfaces, mechatronics engineers create and construct space robotics and manipulation systems. Incorporating state-of-the-art sensors, end-effectors, and robotic limbs, these systems are able to reliably and precisely carry out complex operations in microgravity or dangerous conditions. Reducing the complexity and expense of space missions, space robotics allow for efficient building, repair, and maintenance operations in space.

2.5.4 Precision Guidance and Control Systems

Spacecraft, launch vehicles, and orbital platforms rely heavily on mechatronics engineering for their precise navigation and control systems. To guarantee precise trajectory tracking, orbital maneuvering, and rendezvous operations in space, mechatronics engineers provide guiding algorithms, navigation sensors, and attitude control systems. These systems allow for complex space operations including docking, planetary landing, and orbital rendezvous while optimizing fuel efficiency and reducing mission risks.

2.5.5 Space Habitat and Life Support Systems

In order to ensure that astronauts have a safe and livable environment to live in during long-duration missions, mechatronics engineers help develop and integrate space habitats and life support systems. Intelligent habitat management systems regulate radiation levels, air quality, temperature, and humidity inside spacecraft and habitats. ECLSS and regenerative life support technologies are also part of this system. In order to keep humans in space for long periods of time, mechatronics engineers must guarantee that these systems are autonomous, efficient, and reliable.

Autonomous spacecraft, additive manufacturing, space robotics, precision navigation systems, and space housing technologies are all part of the future trend in mechatronics engineering for space exploration and aerospace. When it comes to exploring and inhabiting space, mechatronics engineers are at the forefront of technological innovation, which is paving the way for game-changing technology.

2.6 *Future Trend of Mechatronics Engineering on Agricultural Automation*

The field of mechatronics engineering has the potential to transform farming by utilizing data-driven technology, robotics, and automation to tackle issues including labor shortages, sustainable agricultural techniques, and the world's food supply. The field of mechatronics engineering within agricultural automation is being influenced by a number of important trends.

2.6.1 Autonomous Farm Machinery

Autonomous tractors, harvesters, and drones outfitted with cutting-edge sensing, navigational, and control systems are being developed by mechatronics engineers. By maximizing resource use, decreasing labor costs, and increasing agricultural yields, these autonomous systems make it possible to plant, water, fertilize, and harvest with pinpoint accuracy. Drones with cameras and sensors also allow farmers to keep tabs on their crops, pest infestations, and environmental conditions in real time, which helps them make better decisions and respond faster.

2.6.2 Robotic Crop Management

Crop management activities including weeding, trimming, and spraying are enhanced by robotic systems that are developed through mechatronics engineering. In order to minimize the usage of herbicides and pesticides, these robotic systems employ computer vision, machine learning, and robotic manipulation techniques to detect and selectively target weeds, pests, and unhealthy plants. In large-scale monoculture and organic farming systems, robotic crop management systems improve crop quality, reduce environmental impact, and enhance agricultural efficiency.

2.6.3 Precision Agriculture and IoT Integration

In order to monitor and regulate farming characteristics including soil moisture, temperature, nutrient levels, and crop growth in real time, mechatronics engineers are combining sensors, actuators, and Internet of Things connectivity into precision agriculture systems. Optimizing irrigation timing, fertilizer delivery, and pest management procedures are achieved by the data collected and analyzed by these systems from a variety of sources, including ground-based sensors, drones, and satellites. By adapting farming methods to individual fields and crops, precision agriculture technologies maximize efficiency, reduce input costs, and promote sustainability (see Fig. 2).

2.6.4 Vertical Farming and Controlled Environment Agriculture

Vertical farming and controlled environment agriculture (CEA) technologies use mechatronics engineering for the design and automation of urban and indoor facilities to produce high value crops all year round. To create the best possible growth conditions with the least amount of water, energy, and impact on the environment, mechatronics engineers design automated hydroponic and aeroponic systems, LED lighting solutions, and climate control systems. A combination of vertical farming and CEA technology can make cities more resilient to climate change by increasing food security, decreasing transportation costs, and allowing for local food production.

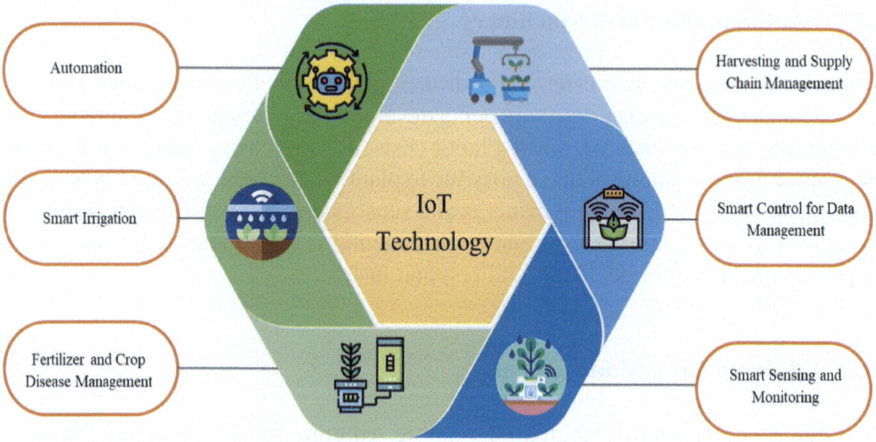

Fig. 2 IoT-based technological advancements in agricultural sector

2.6.5 Data Analytics and Decision Support Systems

Data analytics, machine learning, and decision support systems are being used by mechatronics engineers to sift through massive amounts of agricultural data and deliver practical advice to agronomists, farmers, and others involved in the agricultural industry. With the use of these technologies, farmers can optimize their production plans, make educated decisions, and anticipate crop yields, disease outbreaks, and market trends. To guarantee the sustainability and long-term viability of agricultural systems, data-driven agriculture boosts profitability, productivity, and resistance to environmental unpredictability.

Autonomous farm machinery, precision agriculture, data-driven decision support systems, vertical farming, robotic crop management, and autonomous farm machinery are all parts of the future trend in agricultural automation as engineered by mechatronics engineers. Merging cutting-edge tech with their technical know-how, mechatronics engineers are revolutionizing farming methods, creating a more sustainable and resilient food production future, and driving innovation in the agricultural sector.

2.7 *Future Trend of Mechatronics Engineering on Consumer Electronics and Smart Devices*

Improvements in connection, downsizing, and intelligent functioning are being propelled by mechatronics engineering, which is leading the charge in consumer electronics and smart gadgets. In this area of study, mechatronics engineering is being influenced by a number of important themes (see Fig. 3).

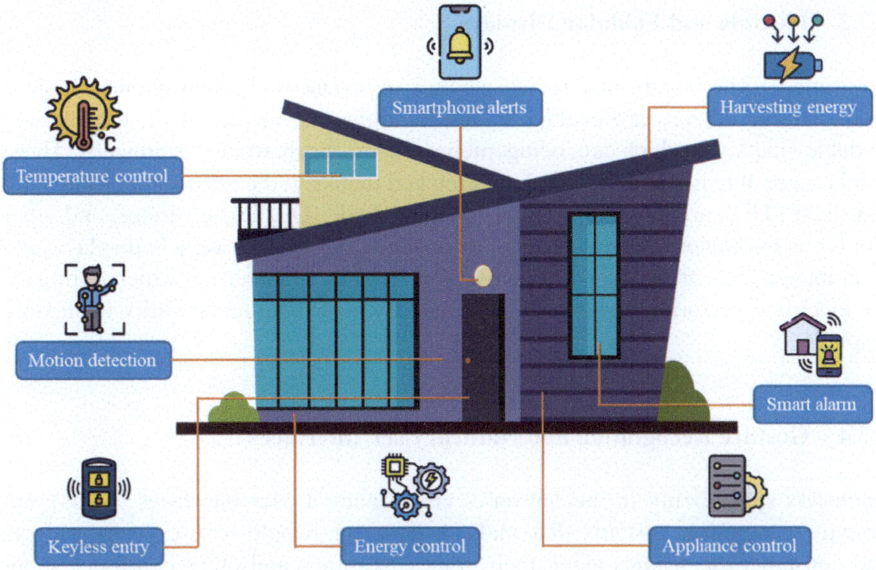

Fig. 3 Day-to-day usage of consumer and smart electronics technologies

2.7.1 Internet of Things (IoT) Integration

To facilitate easy connection and interaction with other devices and Internet services, mechatronics experts are connecting consumer electronics and smart gadgets to the Internet of Things. The ability to remotely monitor, operate, and automate household appliances, wearable technology, and personal gadgets is made possible by these networked devices, which gather and share data over wireless networks. New features like voice control, predictive maintenance, and smart home automation are made possible by the integration of the Internet of Things (IoT), which also improves user convenience, customization, and energy efficiency.

2.7.2 Wearable Technology

Smartwatches, fitness trackers, and augmented reality (AR) glasses are all products of the design and development processes that involve mechatronics engineering. By combining sensors, actuators, and microelectronics, these wearable gadgets are able to keep tabs on users' activity levels, measure vital signs, and provide them with tailored messages and feedback. Wearable technology relies on the work of mechatronics experts who optimize their shape, power consumption, and user interface to make them as practical, comfortable, and easy to use as possible.

2.7.3 Flexible and Foldable Displays

Innovative form factors and increased user experiences in smartphones, tablets, and wearable devices are enabled by technological developments in flexible and foldable displays, which are being pioneered by mechatronics engineers. These displays are able to be bent, rolled, or stretched thanks to the utilization of thin-film transistor (TFT) arrays, organic light-emitting diode (OLED) substrates, and other similar technologies. Innovative uses in consumer electronics, vehicle displays, and wearable gadgets are made possible by mechatronics engineering, which optimizes the electrical performance, mechanical integrity, and manufacturability of flexible displays.

2.7.4 Gesture Recognition and Natural User Interfaces

Engineers specializing in mechatronics create natural user interfaces (NUIs) and gesture recognition systems that make it easy for people to use smart gadgets and consumer electronics using their voices, gestures, and other non-touch input methods. Users are able to operate objects and apps without physically touching them because of these systems' usage of sensors, cameras, and machine learning algorithms that read hand gestures, facial expressions, and spoken instructions. With the use of gesture detection and natural user interfaces (NUIs), users may enjoy VR, entertainment, and gaming experiences that are more accessible, convenient, and immersive.

2.7.5 Energy Harvesting and Sustainable Design

Improvements in energy harvesting, power management, and environmentally friendly materials are some of the ways that mechatronics engineering helps bring about more sustainable and energy-efficient consumer electronics. To lessen their devices' negative effects on the environment, mechatronics experts create rechargeable batteries, low-power sensors, and energy-efficient circuitry. Environmental responsibility and the circular economy are further advanced by the incorporation of sustainable design concepts into the creation of consumer electronics. These ideas include energy-efficient manufacturing techniques, recyclable materials, and modular construction.

Finally, sustainable design, gesture recognition, flexible displays, wearable tech, and the Internet of Things are all parts of the future trend in consumer electronics and smart device mechatronics engineering. In the ever-changing world of consumer electronics and smart gadgets, mechatronics engineers are at the forefront of technical innovation, paving the way for game-changing innovations that improve connection and user expertise.

SPRINGER NATURE

GPSR Compliance

The European Union's (EU) General Product Safety Regulation (GPSR) is a set of rules that requires consumer products to be safe and our obligations to ensure this.

If you have any concerns about our products, you can contact us on ProductSafety@springernature.com

In case Publisher is established outside the EU, the EU authorized representative is:

Springer Nature Customer Service Center GmbH
Europaplatz 3
69115 Heidelberg, Germany

The manufacturer's authorised representative in the EU is Springer Nature Customer Service Centre GmbH, Europaplatz 3, 69115 Heidelberg, Germany. If you have any concerns regarding our products, please contact ProductSafety@springernature.com

Printed and bound by CPI Group (UK) Ltd, Croydon, CR0 4YY
01/10/2026
02225857-0002